Digital integrated circuits and computers

LIBRAR

Accession

Digital integrated circuits and computers

Barry G Woollard C Eng, MIERE, M Inst MC

*Lecturer in Industrial Electronics, Instrumentation
and Control Engineering, Walsall College of Technology*

McGRAW-HILL Book Company (UK) Limited

London · New York · St Louis · San Francisco · Auckland
Beirut · Bogotá · Düsseldorf · Johannesburg · Lisbon
Lucerne · Madrid · Mexico · Montreal · New Delhi
Panama · Paris · São Paulo · Singapore · Sydney · Tokyo
Toronto

Published by
McGRAW-HILL Book Company (UK) Limited
MAIDENHEAD · BERKSHIRE · ENGLAND

British Library Cataloguing in Publication Data

Woollard, Barry
  Digital integrated circuits and computers. —
  (Technician education series).
  1. Digital integrated circuits   2. Electronic
  digital computers
  I. Title   II. Series
  621.381'73'7   TK7874   78-40020

  ISBN 0-07-084233-7

12345 J.W.A. 80798

PRINTED AND BOUND IN GREAT BRITAIN

Contents

Preface

Developments in the technology of integrated circuits during the last decade have caused major changes to be made in the field of digital electronics and computing. These advances have led to the widespread use of electronic calculators and digital watches in the domestic field; and to digital multimeters, digital frequency counters, digital panel meters (DPM's), direct digital control of industrial processes and the micro-computer (together with all its implications) in the industrial field.

Digital integrated circuits are manufactured using two main technologies: *bipolar* techniques, e.g., TTL, and *unipolar* techniques, e.g., MOS, CMOS. The principles of both types are considered, but the applications concentrate on the use of TTL circuits. However, providing the necessary precautions are observed, the same applications may be realised using CMOS circuits.

This text forms the basis of the content of one of a range of very successful short courses developed at Walsall College of Technology, in which the aims have been to encourage as much practical involvement as possible, to enable the student to gain experience in handling, connecting and making measurements on digital integrated circuits. I would like to thank all those who have helped in the preparation of this book, in particular to my colleagues, students and to the many people in industry who have given me advice. Finally, a special thanks to my wife, for her efforts and patience while typing the manuscript.

Barry G. Woollard

1 Integrated circuits

1.1 Introduction

Integrated circuits (IC's) may be broadly categorised into *Linear IC's* which contain amplifying-type circuitry, and *Digital IC's* which contain switching-type circuitry. This book will consider Digital IC's only, i.e., the basic function of these IC's is to handle DIGITAL information by means of switching circuits. Of course, another general use for switching circuits is to control power for industrial machinery, e.g., motors. But, due to the limitations of power handling of IC's, they are not yet widely used for these purposes.

Digital IC's are used to PROCESS information and STORE information in such digital systems as computers, desk and pocket calculators, machine tool controls, and frequency counting instruments. When using IC's, designers no longer need to spend time and effort in putting together transistors to make gates and flip-flops. They do not even have to put together gates and flip-flops to make more complex circuits. Instead they can design a system by using larger building blocks, made up of many gates and flip-flops. Indeed, most of the larger building blocks needed by designers of digital systems are now available in IC form. Therefore, to learn about digital IC's requires becoming familiar with some of the standard building blocks.

1.2 Development of integrated circuits

The development of integrated circuits has followed from the manufacturing techniques evolved for transitors. Early types of transitor were *grown junction*, the p and n regions being grown within the single crystal. These were followed by the *alloy-junction* in which n type impurity pellets are alloyed on opposite sides of a p type wafer (giving a base width of approximately 25 μm), as shown in Fig. 1.1. This gave improved high frequency performance, but only to a maximum of about 1 MHz. *Alloy-diffused* transistors, having very thin base regions, as shown in Fig. 1.2, were developed to improve performance—up to several hundred megahertz— but these had limited collector-to-emitter voltage ratings.

The *planar* process, which is now virtually the only process for silicon devices, was first described in 1961. Very high frequency performances—

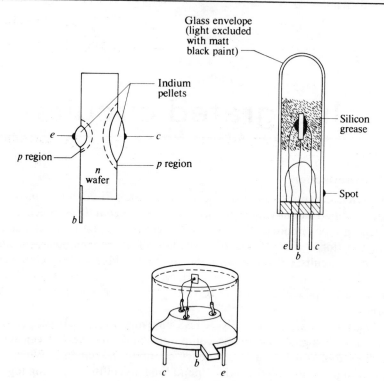

Fig. 1.1. The alloy-junction transistor.

several thousand megahertz—can be achieved, and several thousand transistors can be manufactured on one 'slice' of material; but for simplicity, one transistor only will be considered. An n type slice of semiconductor material has an oxide layer deposited over its surface. 'Windows' are then etched in the silicon oxide by a photo-resist process as shown in Fig. 1.3. The slice is then exposed to a p type atmosphere and the base region is diffused into the n type slice through the 'window' as shown in Fig. 1.4. The

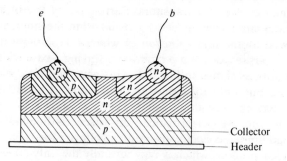

Fig. 1.2. The alloy-diffused transistor.

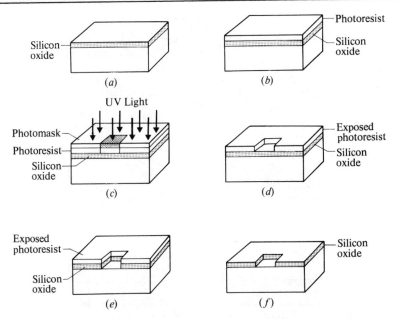

Fig. 1.3. The photo-resist process. (*a*) Oxidized silicon slice, (*b*) photo-resist lacquer applied to surface, (*b*) photo-resist exposed to uv through photo-mask, (*d*) unexposed photo-resist removed with solvent, (*e*) silicon oxide in 'window' removed by etching, (*f*) photo-resist removed.

surface is re-oxidized, etched, and the emitter n region diffused into the base region as shown in Fig. 1.4. The masks used in the photo-resist sequence are as shown in Fig. 1.5 for the one transistor. The exposed edges of p–n junctions are formed underneath previously deposited oxide layers. This provides the protection from the environment which the p–n junction needs.

In order to obtain a low collector capacitance and a high breakdown voltage it is desirable for the collector region to have a high resistivity. However, a high resistivity produces a significant and undesirable resistance between the collector junction and the collector contact. This problem is overcome by the *epitaxial* process in which a thin layer of high resistivity n material is deposited on top of a thick substrate of low resistivity n^+ material. Diffusion is a useful process for creating layers of low resistivity on high resistivity substrates but it does not work satisfactorily the other way round. The epitaxial process is a vapour deposition of high purity semiconductor on a lower purity substrate which nevertheless continues the crystal structure of the substrate. Now, if the transitor process is started with a low resistivity n^+ substrate upon which a high resistivity n layer has been epitaxially deposited, the diffusion processes can take place in the epitaxial layer only, thus achieving the low capacitance and high breakdown voltage. But the necessarily thick substrate does not

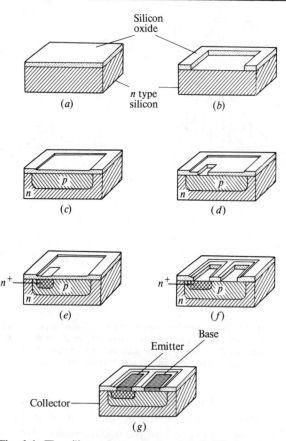

Fig. 1.4. The silicon planar process. (*a*) **Oxidized *n* type silicon, (*b*) base window cut in oxide by first photo-resist stage, (*c*) *p* type base region diffused and new oxide formed, (*d*) emitter window cut by second photo-resist stage, (*e*) *n*+ emitter diffused and new oxide formed, (*f*) base and emitter contact window cut by third photo-resist stage, (*g*) aluminium contacts evaporated and then defined by fourth photo-resist stage.**

now have a high resistance. The planar transistor is then formed in the epitaxial layer to form the *silicon planar epitaxial* transistor as shown in Fig. 1.6.

Complete circuits—*monolithic integrated circuits* are constructed by these methods—in which resistors, capacitors, diodes and transitors are laid down in a common epitaxial layer, and connected to one another by metallic interconnections, which are evaporated on to the oxide layer, as shown in Fig. 1.7. These are generally referred to as *bipolar techniques*—since fabrication includes both p and n type semiconductor material. The sequence of fabrication of an integrated circuit is illustrated in Fig. 1.8, and a typical integrated circuit is shown in Fig. 1.9.

4

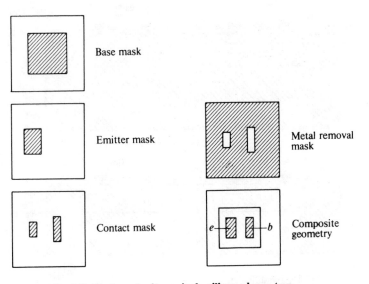

Fig. 1.5. Photomasks for a single silicon planar transistor.

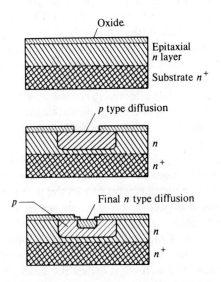

Fig. 1.6. The silicon planar epitaxial transistor.

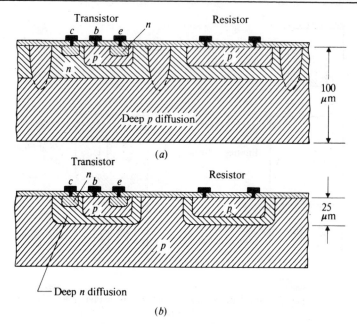

Fig. 1.7. First generation integrated circuits. (*a*) Deep diffused process, (*b*) triple diffused process.

More recently, *unipolar* transistors have been developed. They are unipolar due to current being conveyed through them by majority charge carriers only. These transistors depend for their action upon the effect of an electric field, and are thus called *field effect transistors* (FET's).

A simple n channel junction gate FET (JUGFET or JFET) consists of a slice of n type semiconductor (the *channel*) with a p type region formed in it (the *gate*) as shown in Fig. 1.10. The majority charge carriers flow from the *source* (*emitter*) to the *drain* (*collector*). A reverse bias applied to the p–n junction forms a depletion layer at the junction as shown in Fig. 1.11 (*a*). The depletion layer effectively reduces the width of the channel—thus increasing its resistance and reducing the drain current. Increasing the gate voltage increases the width of the depletion layer which reduces the drain current further.

Alternatively, if the gate voltage is zero, and the drain voltage is increased, a current is caused to flow longitudinally in the channel, and the p.d. along the face of the gate region causes the p–n junction to be reverse biased thus forming a depletion layer as shown in Fig. 1.11 (*b*). Further increase in drain (or gate) voltage eventually causes the depletion layer to extend across the width of the channel to 'pinch off' the current into a very thin layer (whose resistance is about 250 kΩ). At zero gate voltage, the value of drain voltage at pinch off is called the pinch off voltage. At drain voltages above pinch off, the output characteristic indicates saturation.

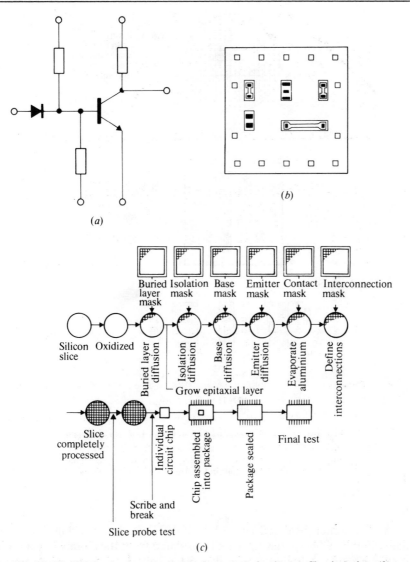

Fig. 1.8. The fabrication sequence of an integrated circuit. (*a*) Circuit design, (*b*) circuit layout, (*c*) photomask fabrication (i.e., artwork, photographic reduction, step and repeat).

This operation is described as the *depletion mode*—since an increase in reverse bias signal at the gate causes the channel to be depleted. A JFET can be operated with a very small forward biased voltage at the gate which reduces the depletion layer and increases the drain current—this is the *enhancement mode*. The main advantage of the FET over the bipolar transistor is that it has a very high input impedance (hundreds of megohms) since the gate is not designed to carry current.

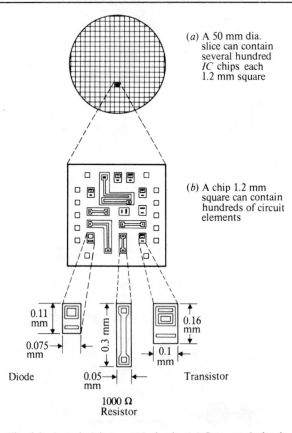

(a) A 50 mm dia. slice can contain several hundred IC chips each 1.2 mm square

(b) A chip 1.2 mm square can contain hundreds of circuit elements

Fig. 1.9. A typical integrated circuit. (a) Integrated circuit slice. A 50 mm diameter slice can contain several hundred IC chips each 1.2 mm square, (b) integrated circuit chip. A chip 1.2 mm square can contain hundreds of circuit elements.

An n channel insulated gate FET (IGFET) is shown in Fig. 1.12, the main difference being that the gate is insulated from the channel by a layer of silicon oxide. For this reason it is often called the *metal oxide silicon* transistor (MOS transistor).

The IGFET is most often operated in the enhancement mode. A positive potential applied to the gate attracts many minority charge carriers to the semiconductor-oxide interface—which causes an *inversion channel* (n type material) to be formed. The higher the positive potential the greater is the conductivity of the inversion channel, i.e., increase in gate voltage *enhances* the current flow in the channel between source and drain.

Field effect transistors have led to the development of complementary MOS (CMOS) integrated circuits. These will be briefly discussed in chapter 3.

8

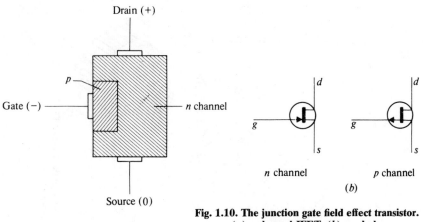

Drain (+)

Gate (−)

p

n channel

Source (0)

(a)

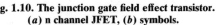

n channel

p channel

(b)

Fig. 1.10. The junction gate field effect transistor.
(a) n channel JFET, (b) symbols.

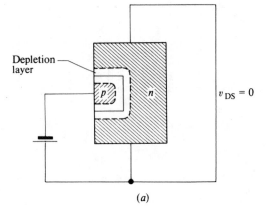

Depletion layer

p

n

$v_{DS} = 0$

(a)

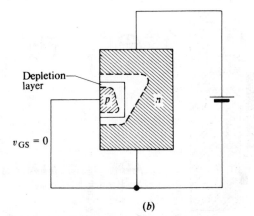

Depletion layer

p

n

$v_{GS} = 0$

(b)

Fig. 1.11. Formation of depletion layer in JFET.
(a) Reverse bias applied to gate, (b) longitudinal
electric field.

9

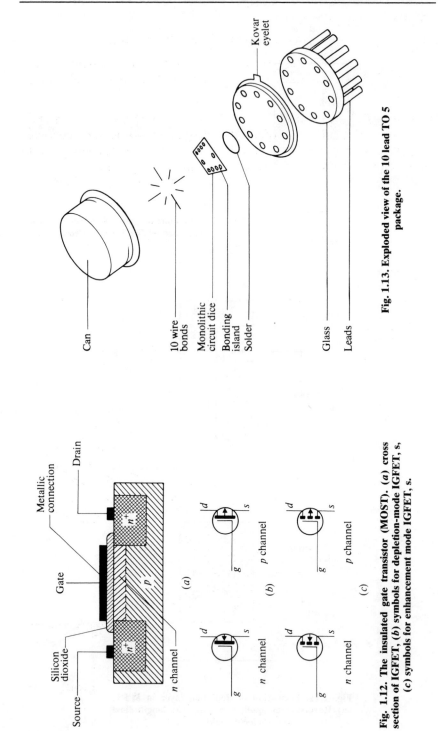

Fig. 1.13. Exploded view of the 10 lead TO 5 package.

Fig. 1.12. The insulated gate transistor (MOST). (a) cross section of IGFET, (b) symbols for depletion-mode IGFET, s, (c) symbols for enhancement mode IGFET, s.

1.3 Packaging of IC's

Three packages are currently in use for integrated circuits:

(a) *TO 5 package.* This is a reduced height version of the popular transistor outline TO 5 can and is available in 8 or 10 lead versions, as shown in Fig. 1.13.

(b) *Flat pack (14 lead).* This package was designed to be welded to printed circuit boards and this obviates the need for unreliable soldered joints. It was also born out of the need for greater flexibility of connections to the outside world, i.e., more leads, as shown in Fig. 1.14.

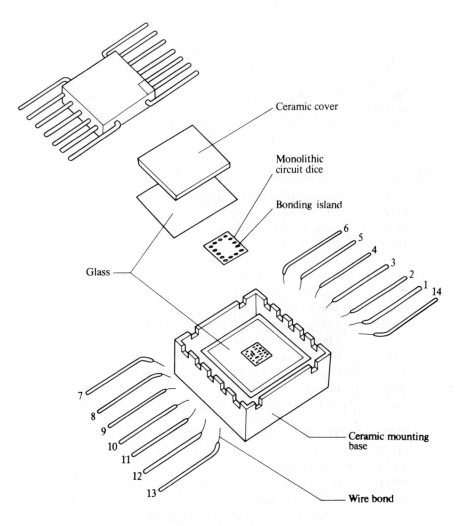

Fig. 1.14. 14 lead flat package.

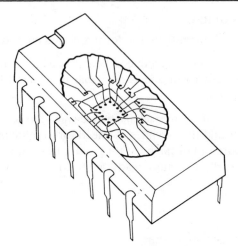

Fig. 1.15. DIL package.

(c) *DIL plastic package*. This dual-in-line package is intended as a cheap multi-lead arrangement as shown in Fig. 1.15. Variations of this include 16, 20, 24 and 40 pin arrangements.

1.4 Abbreviations commonly encountered

BJT	Bipolar junction transistor
UJT	Unijunction transistor
LCD	Liquid crystal display
LED	Light emitting diode
FET	Field effect transistor
JFET	Junction gate FET
IGFET	Insulated gate FET
MOS	Metal oxide silicon
CMOS	Complementary MOS (COSMOS)
LOCMOS	Locally oxidized MOS
DIL	Dual-in line
DRL	Diode resistor logic
RTL	Resistor transistor logic
DTL	Diode transistor logic
ECL	Emitter coupled logic
TTL	Transistor transistor logic (T^2L)
HTTL	High speed TTL
LTTL	Low power TTL
STTL	Schottky TTL
IIL	Integrated injection logic (I^2L)
SSI	Small scale integration (<10 gates)
MSI	Medium scale integration ($10-100$ gates)

LSI	Large scale integration (>100 gates)
RAM	Random access memory
ROM	Read only memory
PROM	Programmable ROM
EPROM	Erasable PROM
EAROM	Electrically alterable ROM
CAM	Content addressable memory
CPU	Central processor unit
ALU	Arithmetic logic unit
SCR	Sequence control register
NC	Not connected (no connection)
PISO	Parallel-in, serial-out
PIPO	Parallel-in, parallel-out
SIPO	Serial-in, parallel-out
FIFO	First-in, first-out
LIFO	Last-in, first-out

2 Logic, basic logic gates and numbering systems

2.1 Logic

The application of logic in this context refers to control, alarm and monitoring systems on a relatively simple scale, and its use in more complex systems for data processing and computer control can be implied from this. Logic systems may be divided into two main groups:

(a) *Combinational systems* have provision for a combination of inputs, to which one or more outputs respond. If a particular combination of input signals are applied to the system, an output condition will occur in the same instant and will persist for as long as the input condition is applied.

(b) *Sequential systems* the output condition at any instant depends on the nature of combinations of input signals which have been received in sequence up to that instant. Sequential systems may be very similar to combinational systems in design and the elements used, but also have facilities for memorising past occurrences at the inputs.

These two systems may be further classified into:

(c) *Synchronous systems* in which the outputs are prevented from changing to correspond with a given input arrangement until a timing signal is received. This timing signal is fed to all parts to ensure that each part operates in synchronism with the rest. The nature of the timing signal is a train of pulses (*clock pulses*) which are generated independently of the system by a master pulse generator or 'clock'.

(d) *Asynchronous systems* in which the various parts are not synchronised by clock pulses, but it is required that the various outputs hold their states until they have been sensed by the rest of the system.

2.2 Boolean algebra

In 1854 George Boole wrote a paper 'An Investigation into the Laws of Thought'. Normal mathematics, although extremely useful in many intellectual pursuits, cannot deal with every aspect of thought.

14

The following example shows the inadequate nature of normal algebra. Suppose the following statements are made:

Cats are animals.

Dogs are animals.

Therefore cats are dogs.

The conclusion is obviously absurd!

However, suppose we represent the statements in normal algebra:

A = B

C = B

∴ A = C which is a perfectly valid deduction.

The inherent difficulty is a question of LANGUAGE. The mathematical sign of equality is used to represent the word 'are'. The fact that a cat is an animal is true, but it is not true to say that it 'equals' an animal, since it is only a SUB-class of the much larger general class.

Boole developed an entirely new system and called it 'The Algebra of Classes'. Apart from its interest to the mathematical genii of the age, Boole's treatise rotted away in the corners of the world's libraries until 1938. By that time telephone and communication engineering had reached a high degree of complexity and Boole's methods were suddenly famous again, due to a paper entitled 'Symbolic Analysis of Relay and Switching Circuits' published by C. E. Shannon.

Shannon discovered that Boole's Algebra of Classes was a powerful tool with which to analyse and represent complicated circuitry employing 'two-state' ideas.

2.3 Basic rules of Boolean algebra

(a) A quantity can have only one of two possible values, it can be a 1 or a 0. No other value exists.

(b) The usual meaning of certain mathematical signs takes on an entirely different meaning:

A.B means A *AND* B, not A times B

A + B means A *OR* B, not A and B

$\bar{A}$ means NOT A (or, the complement of A)

(c) The sign of equality (=) has a new significance, and may best be defined as follows:

= means 'an output exists', or 'the switch is closed'.

2.4 The laws of logic

The truth of many logic statements is self-evident, whilst that of others may not be so clear. Providing the statement is accurate, it is possible to test its truth.

Using binary notation, we say that a statement is true, i.e., the function exists, if it has a value 1. If it is false, it does not exist, and has a value 0.

15

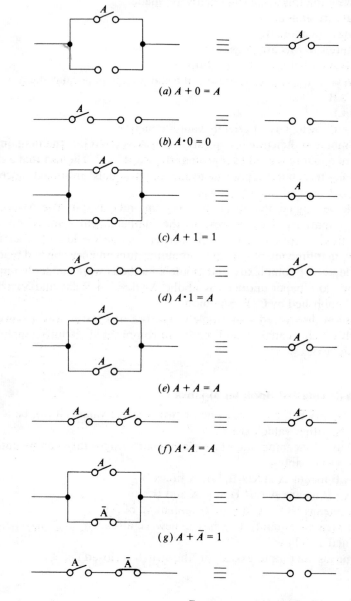

(a) $A + 0 = A$

(b) $A \cdot 0 = 0$

(c) $A + 1 = 1$

(d) $A \cdot 1 = A$

(e) $A + A = A$

(f) $A \cdot A = A$

(g) $A + \bar{A} = 1$

(h) $A \cdot \bar{A} = 0$

Fig. 2.1. The laws of logic.

(a) $A+0 \quad = A$
(b) $A.0 \quad\quad = 0$
(c) $A+1 \quad = 1$
(d) $A.1 \quad\quad = A$
(e) $A+A \quad = A$
(f) $A.A \quad\quad = A$
(g) $A+\bar{A} \quad = 1$
(h) $A.\bar{A} \quad\quad = 0$
(i) $\overline{A+B+C} = \bar{A}.\bar{B}.\bar{C}$
(j) $\overline{A.B.C} \quad = \bar{A}+\bar{B}+\bar{C}$

The first eight of these laws may be explained by examining simple switching arrangements as shown in Fig. 2.1, in which a normally open switch is designated A, and a normally closed switch is $\bar{A}$.

2.5 Symbolic representation of logic elements

The graphical symbols used in pure logic diagrams represent thought processes, and are therefore independent of the equipment which might be used to implement them, i.e., the same symbols are common to all methods of implementation, whether it be electronic, pneumatic, hydraulic, mechanical, etc.,—although, during this text we shall only be dealing with electronic equipment.

If all signal lines in a system have the same pair of physical states, and if both are electric potentials (or currents), then if the more positive potential is consistently selected as the 1 state the resultant system uses POSITIVE LOGIC.

If the less positive potential is consistently selected as the 1 state, the resultant system uses NEGATIVE LOGIC.

If neither positive nor negative potentials are consistently selected as the 1 state, the resultant system has MIXED LOGIC.

2.6 Basic logic gates and truth tables

(a) *The AND gate.* The logic symbol is as shown in Fig. 2.2 (*a*). The output F is 1 when the input signals at A *AND* B *AND* C are at 1. The truth table gives the output state for all possible combinations of inputs as shown in Fig. 2.2 (*b*).

      The Boolean equation for the AND gate is therefore:

$$F = A.B.C$$

(b) *The OR gate.* The logic symbol is as shown in Fig. 2.3 (*a*). The output F is 1 when the input signal at A *OR* B *OR* C *OR* ANY *OR* ALL are 1 as shown in the truth table in Fig. 2.3. (*b*).

      The Boolean equation for the OR gate is therefore:

$$F = A+B+C$$

17

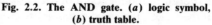

(a)

A	B	C	F
0	0	0	0
0	0	1	0
0	1	0	0
0	1	1	0
1	0	0	0
1	0	1	0
1	1	0	0
1	1	1	1

(b)

**Fig. 2.2. The AND gate. (a) logic symbol,
(b) truth table.**

(c) *The NOT gate.* (Inverter, Negater). The logic symbol for this gate is as shown in Fig. 2.4 (a). Note the small circle indicating the inhibiting action. Alternatively, the symbol may be as shown in Fig. 2.4 (b). The output signal represents 1 when the input signal represents 0, and vice-versa, i.e., the input signal is complemented.

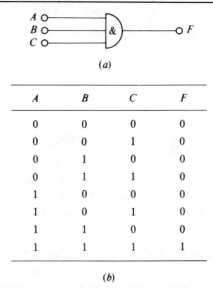

(a)

A	B	C	F
0	0	0	0
0	0	1	1
0	1	0	1
0	1	1	1
1	0	0	1
1	0	1	1
1	1	0	1
1	1	1	1

(b)

**Fig. 2.3. The OR gate. (a) logic symbol,
(b) truth table.**

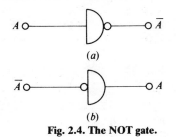

(a)

(b)

Fig. 2.4. The NOT gate.

(d) *The NAND gate.* (NOT-AND). The logic symbol is as shown in Fig. 2.5 (*a*). The output F is 0 when the input signals at A *AND* B *AND* C are 1 as shown in the truth table in Fig. 2.5 (*b*).

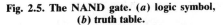

(a)

A	B	C	F
0	0	0	1
0	0	1	1
0	1	0	1
0	1	1	1
1	0	0	1
1	0	1	1
1	1	0	1
1	1	1	0

(b)

**Fig. 2.5. The NAND gate. (*a*) logic symbol,
(*b*) truth table.**

The Boolean equation for the NAND gate is therefore:

$$F = \overline{A.B.C}$$

(e) *The NOR gate* (NOT-OR). The logic symbol is as shown in Fig. 2.6 (*a*). The output F is 0 when ONE or MORE of the input signals is 1 as shown in the truth table in Fig. 2.6 (*b*).

The Boolean equation for the NOR gate is therefore:

$$F = \overline{A + B + C}$$

Note: A positive logic NAND gate is a NEGATIVE logic NOR gate, and vice versa. Verify this by comparing the Truth Tables.

19

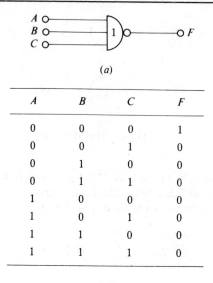

(a)

A	B	C	F
0	0	0	1
0	0	1	0
0	1	0	0
0	1	1	0
1	0	0	0
1	0	1	0
1	1	0	0
1	1	1	0

(b)

Fig. 2.6. The NOR gate. (a) logic symbol, (b) truth table.

2.7 Numbering systems

(a) *Denary.* Use TEN symbols representing the quantities 0 through to 9. The number of digits used in the system is known as its BASE (or RADIX), i.e., TEN in this case.

Other numbers are *constructed* by giving different values, or *weights,* to the position of the digit relative to the denary (decimal) point. The weights of the different positions are given by powers of the radix, and in general is:

$$R^2 \quad R^1 \quad R^0 \quad \cdot \quad R^{-1} \quad R^{-2}$$

which, for the denary system is:

$$10^2 \quad 10^1 \quad 10^0 \quad \cdot \quad 10^{-1} \quad 10^{-2}$$

Example

$$426_{10} = 4 \times 10^2 + 2 \times 10^1 + 6 \times 10^0$$

$$= 400 + 20 + 6$$

$$= 426$$

(b) *Octal.* Widely used in computer programming. Easy to convert to binary and vice-versa. Octal is easier to work with than binary due to the reduced number of digits.

Radix is 8, and positional weights are powers of 8:

$$8^2 \quad 8^1 \quad 8^0 \quad \cdot \quad 8^{-1} \quad 8^{-2}$$

Example

$$426_8 = 4 \times 8^2 + 2 \times 8^1 + 6 \times 8^0$$
$$= 4 \times 64 + 2 \times 8 + 6 \times 1$$
$$= 256 + 16 + 6$$
$$= 278_{10}$$

Denary		Octal		Binary				
Tens 10^1	Ones 10^0	Eights 8^1	Ones 8^0	Sixteens 2^4	Eights 2^3	Fours 2^2	Twos 2^1	Ones 2^0
	0		0					0
	1		1					1
	2		2				1	0
	3		3				1	1
	4		4			1	0	0
	5		5			1	0	1
	6		6			1	1	0
	7		7			1	1	1
	8	1	0		1	0	0	0
	9	1	1		1	0	0	1
1	0	1	2		1	0	1	0
1	1	1	3		1	0	1	1
1	2	1	4		1	1	0	0
1	3	1	5		1	1	0	1
1	4	1	6		1	1	1	0
1	5	1	7		1	1	1	1
1	6	2	0	1	0	0	0	0
1	7	2	1	1	0	0	0	1
1	8	2	2	1	0	0	1	0
1	9	2	3	1	0	0	1	1
2	0	2	4	1	0	1	0	0
2	1	2	5	1	0	1	0	1
2	2	2	6	1	0	1	1	0
2	3	2	7	1	0	1	1	1
2	4	3	0	1	1	0	0	0
2	5	3	1	1	1	0	0	1

Fig. 2.7. Comparison between numbering systems.

(c) *Binary.* Very widely used in logic and computing. Radix is 2, and positional weights are powers of 2:

$$2^4 \quad 2^3 \quad 2^2 \quad 2^1 \quad 2^0 \quad \cdot \quad 2^{-1} \quad 2^{-2} \quad 2^{-3}$$

Several rules exist for dealing with addition, subtraction, multiplication and division, and for converting between one numbering system and another, some of which will be further discussed in chapter 8.

A comparison between the above numbering systems is shown in Fig. 2.7.

(d) *Hexadecimal.* Widely used in alpha-numeric display systems and in micro-computing.

In this numbering system a 4-bit binary number is used to represent the numbers 0 to 9 using a conventional count, and the conventional numbers 10 to 15 are represented by the letters A to F as shown in Fig. 2.8.

Denary	Binary				Hexadecimal
0	0	0	0	0	0
1	0	0	0	1	1
2	0	0	1	0	2
3	0	0	1	1	3
4	0	1	0	0	4
5	0	1	0	1	5
6	0	1	1	0	6
7	0	1	1	1	7
8	1	0	0	0	8
9	1	0	0	1	9
10	1	0	1	0	A
11	1	0	1	1	B
12	1	1	0	0	C
13	1	1	0	1	D
14	1	1	1	0	E
15	1	1	1	1	F

Fig. 2.8. Hexadecimal numbering system.

Larger 'hex' numbers may be constructed by using a 4-bit binary group for each hex digit, e.g., 0111 1001 would represent 79. Similarly, the sixteen bit number 1011011011011111 would represent the hex number B6 DF. This is therefore a useful system to interpret.

3 Digital integrated circuits

3.1 Introduction

Electronic logic elements have evolved through a number of stages, beginning with systems consisting of diode AND and OR gates. Advances in semiconductor technology fostered rapid developments in electronic logic circuitry of the active type, and various circuits were produced. The first integrated logic elements were simply translations of discrete component circuits directly into silicon circuits. The earliest types were, in fact composed of several silicon chips with wire interconnections. As integrated circuit techniques developed the design approach changed and the circuits began to be designed to suit the manufacturing technology instead of being duplicates of discrete component prototypes. Once it was realised that circuit complexity was not a limiting factor, the way was open to the production of high-performance, complex circuit elements.

3.2 Choice of logic family

The choice of a logic family for a particular application is generally determined by consideration of the following factors:
- (a) speed of operation,
- (b) noise immunity,
- (c) power dissipation,
- (d) operating temperature range,
- (e) type of package,
- (f) cost,
- (g) availability.

These factors are not necessarily given in any particular order of priority, the individual application generally dictates in which order the various factors must be considered.

3.3 Speed of operation

The speed of a logic gate is defined by its propagation delay, i.e., the time taken for a logic signal to pass through the gate from input to output. One of the contributing factors for this propagation delay may be illustrated by considering a diode, in which the applied voltage is changed from forward to reverse bias. The current in the forward direction does not in fact fall to zero (or to the leakage current value) immediately, since the charge carriers must first recombine with their parent atoms—and hence disappear. This causes a pulse of reverse current to flow as shown in Fig. 3.1, and this takes time to decay to the leakage value. During this time, the reverse current flow causes an electric charge to be stored in the diode junction. Once the charge carriers have been swept away from the junction (storage time), the reverse current decays to the leakage value (transition time). The total delay (recovery time) represents the propagation delay of a signal switched by a diode.

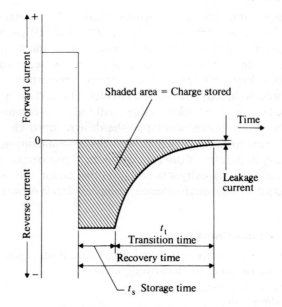

Fig. 3.1. Charge storage in the diode.

Efforts to improve the switching performance of bipolar devices led to the development of the Schottky diode, in which a barrier is formed between a metal and n type semiconductor. The current flow in these devices does not depend in any way on minority charge carriers so that the effects of charge storage are almost eliminated. Schottky junctions are used in diodes and transistors where propagation delays of the order of fractions of nanoseconds are required. The circuit symbols most widely used in the industry for a Schottky diode and transistor are shown in Fig. 3.2.

Some digital equipment, e.g., machine tool control systems, operates at a relatively slow rate where a propagation delay of 1 ms may be acceptable. However, modern high speed digital computers may require propagation

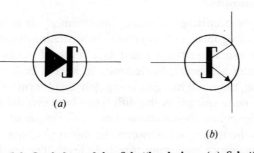

Fig. 3.2. Symbols used for Schottky devices. (a) Schottky diode, (b) Schottky transistor.

delays approaching 1 ns. Most IC logic families manufactured today have propagation delays between about 2 ns and 100 ns.

The asymmetrical switching characteristics of IC's is such that a low to high logic transition signal at the gate input has a different delay to a high to low logic transition at the gate input. These logic levels are referred to as the THRESHOLD LEVELS and specifications generally include a minimum high threshold level and a maximum low threshold level for a particular logic family. Typical propagation delays are usually taken as the average of the two delays stated above, as shown in Fig. 3.3 for an inverter, and is specified at the 50% signal levels.

Stray capacitance at the gate output has a considerable effect on the propagation delay, so that quoted figures generally refer to a given value of capacitance, e.g., 15 pF to 30 pF.

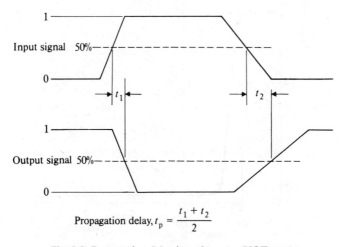

Fig. 3.3. Propagation delay in an inverter (NOT gate).

3.4 Noise immunity

Spurious voltages occurring on signal interconnection paths are termed *noise*. This type of signal can cause erroneous switching of gates. Within the system, noise is usually self-generated as a result of 'cross talk' between signal paths. Gates are generally designed to have a built-in immunity to this type of noise, the noise margins being defined in terms of the threshold levels. The *low noise margin* is the difference between the maximum low output voltage level and the minimum low input threshold level. The *high noise margin* is the difference between the minimum high output voltage level and the maximum high input threshold level. The smaller of the two values is most generally quoted. Thus:

> Noise immunity is the degree to which a gate can withstand variations in input levels without causing a significant change in output state, i.e., the *d.c. noise margin* is *the difference between the output voltage and the input threshold.*

Consider the transfer characteristic of a TTL NAND gate shown in Fig. 3.4. The shaded areas represent forbidden values by the specification, which states that *the output of a gate is guaranteed to be less than 0.4 V in the 0 state, and guaranteed to be greater than 2.4 V in the 1 state. The threshold is guaranteed to be between 0.8 V and 2.0 V.*

Therefore, the *worst-case* noise margins for the TTL NAND gate are:

Noise margin in the 0 state $= 0.8 - 0.4 = 0.4$ V

and, Noise margin in the 1 state $= 2.4 - 2.0 = 0.4$ V

∴ The guaranteed noise margin $= 0.4$ V $= 400$ mV

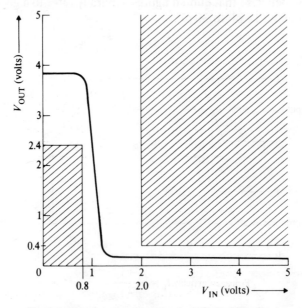

Fig. 3.4. Transfer characteristic of a TTL NAND gate.

3.5   Fan-in and fan-out

The *fan-in* of a logic gate is the maximum number of separate inputs which may be applied to the gate. Fan-in expanders may be used to increase this figure in some cases, but the maximum number is largely determined by the propagation delay of the gate.

The *fan-out* of a logic gate is the maximum number of basic gate inputs that the gate may supply simultaneously without causing the output logic level to fall outside its specification.

3.6   Power dissipation

This is necessary to determine the power supply requirements. Most logic gates draw a different current from the supply depending on whether the output is at logic 1 or 0. Typical values quoted will be the average of the two.

The faster circuits tend to dissipate more power, since they are generally designed with lower values of resistors, also at faster switching speeds, the charging action of the stray capacitance tends to draw more current. Most IC families operate with supply voltages in the region of 5 V, and typical power dissipations range from 1 mW per gate to 100 mW gate.

3.7   Operating temperature range

The operating temperature is the ambient temperature in which the device will operate satisfactorily and meet its specification. Two standard ranges are widely used; the *military* (−55°C to +125°C), and the *commercial* (0°C to +70°C). Certain families are only available in restricted temperature ranges.

3.8   Logic families

Unfortunately, integrated circuit manufacturers tend to introduce their own designs of logic circuit with little standardisation in mind. Eventually however, certain types emerge as being more popular and these are duplicated by other manufacturers, thus providing multiple 'source' availability.

It is usual to classify IC logic families by the circuit configuration of the basic gate function. The main types are:
  (a) Resistor transistor logic (RTL).
  (b) Diode transistor logic (DTL).
  (c) Transistor transistor logic (TTL).
  (d) Emitter coupled logic (ECL).
  (e) CMOS logic.

3.9   Diode resistor logic (DRL)

(a) *Diode AND gate.* Consider the circuit shown in Fig. 3.5 (*a*). When either or both input signals at A and B are connected to logic 0 (0V)

27

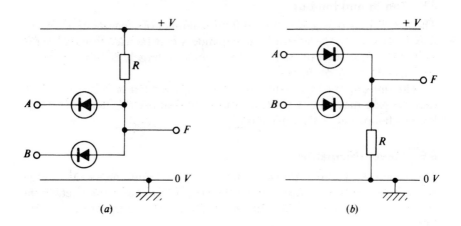

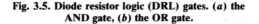

Fig. 3.5. Diode resistor logic (DRL) gates. (a) the AND gate, (b) the OR gate.

the input diode(s) are forward biased and conduct. Under these conditions the current flowing through the resistor R develops a voltage drop, which causes the output signal at F to be low (logic 0). However, this low output is *not* equal to 0V, but *is* equal to the forward voltage drop across the diode, i.e. $V_F \approx 0.7$ V for silicon, and $V_F \approx 0.3$ V for germanium.

Now, when both input signals at A *AND* B are connected to logic 1 (+V volts), the diodes are reverse biased and will therefore not conduct (except for a very small leakage current). Under these conditions the resistor R simply acts as a d.c. connection between output F and the +V supply rail, and the output signal at F is therefore at logic 1 (+V volts).

(b) *Diode OR gate.* Consider the circuit shown in Fig. 3.5 (b). When either or both input signals at A or B are connected to logic 0 (0V) the input diode(s) are reverse biased and do not conduct. Under these conditions no current flows through the resistor R, and the output signal at F is therefore at logic 0 (0V).

When either or both input signals at A *OR* B is connected to logic 1 (+V volts), that diode is forward biased and conducts. Under these conditions the current flowing through R develops a voltage drop, and the output signal at F rises to logic 1. The logic 1 level will be equal to $(+V - V_F)$ volts, which for silicon diodes will be about (V−0.7) volts, and for germanium (V−0.3) volts.

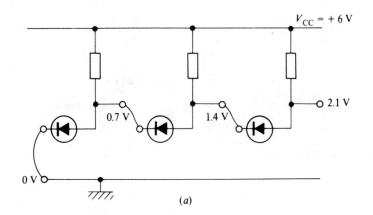

(a)

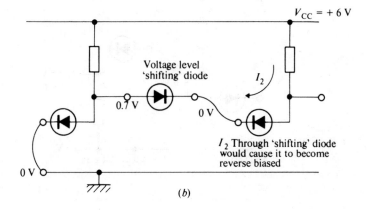

(b)

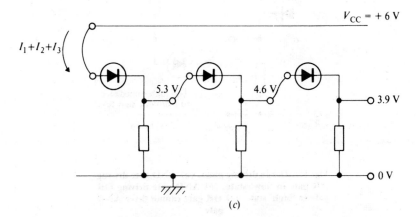

(c)

Fig. 3.6. Cascaded diode gates. (a) Cascaded AND gates, (b) AND gate with 'shifting' diode, (c) cascaded OR gate.

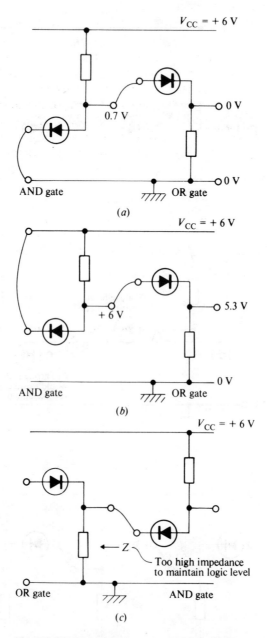

Fig. 3.7. Diode driving gates. (a) AND gate driving OR gate in 'low' state, (b) AND gate driving OR gate in 'high' state, (c) OR gate cannot drive AND gate.

3.10 Cascaded diode gates

When diode AND gates are cascaded, it is necessary to ensure that the voltage shift in the final output (due to the combined effects of the forward voltage drops across the diodes) does not cause erroneous results as shown in Fig. 3.6 (*a*). A forward biased *shifting* diode may be used to remove the voltage shift, but, it is not then possible to connect any further AND gates to the output since the current through the diode of the added gate (in the 'low' state) would reverse bias the 'shifting' diode as shown in Fig. 3.6 (*b*).

When diode OR gates are cascaded the logic 1 signal at each stage is reduced by the forward voltage drop across the diode. In this case the input signal source must also supply current to all the cascaded gates as shown in Fig. 3.6 (*c*).

Diode AND gates may be used to drive diode OR gates as shown in Fig. 3.7 (*a*) and (*b*), but it is not possible for a diode OR gate to drive a diode AND gate since the impedance of the OR gate is too low in the logic 0 state to enable the logic level to be maintained as shown in Fig. 3.7 (*c*).

Practical exercise 3A

DIODE RESISTOR LOGIC

Connect up the circuit arrangements shown in Fig. 3.8.

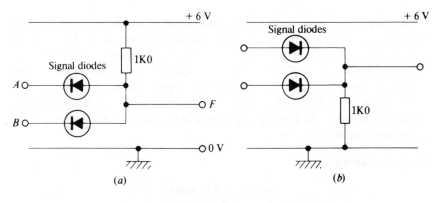

Fig. 3.8. (*a*) Diode AND gate, (*b*) diode OR gate.

Apply all combinations of input signals to A and B and draw up the truth table for both gates. Investigate the *quality* of the logic 1 and logic 0.

Investigate the effects of loading each gate output with further gates to determine the fan-out (about 6 with silicon diodes).

Connect the arrangements shown in Fig. 3.7 to investigate the driving capabilities of each gate.

3.11 Resistor transistor logic (RTL)

This was the first type of logic circuit to be made in integrated circuit form. In the early stages, manufacturer's naturally chose to use well established discrete component circuits on which to base their IC's. A simple RTL NOR gate is shown in Fig. 3.9.

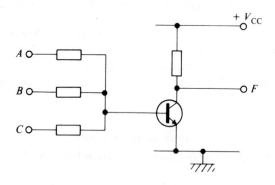

Fig. 3.9. RTL NOR gate.

It is desirable that circuits to be produced in integrated circuit form should have as few passive components as possible. RTL has a large number or resistors compared to the number of transistors. Other disadvantages include low fan-out, low noise immunity—typically 300 mV. This type of gate has a reasonably high speed for a given power dissipation, e.g., 12 to 40 ns at 20 to 2 mW per gate, and forms the basis of the *Mullard NORBIT* system, which is a form of hybrid integrated circuit widely used in industrial control systems.

Practical exercise 3B

DISCRETE RTL GATE

Connect up the discrete circuit shown in Fig. 3.10.

Apply all combinations of input signals and construct the truth table for this circuit. Investigate the *quality* of the logic 1 and logic 0.

Measure the size of a D.U. (drive unit) of current with the components shown.

Investigate the effect of loading the output with other gates. If it is assumed that the minimum high level at the output is 18 V, determine the fan-out of this gate.

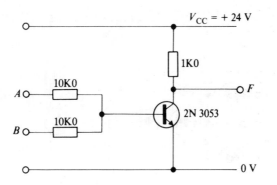

Fig. 3.10. Discrete RTL NOR gate.

How could this value of fan-out be increased?

Note: With 10 K0 input resistors, the D.U. of current is clearly more than enough to saturate the transistor ($V_{CEsat} \approx 0.2$ V for silicon transistors).

When driving a similar gate, the output feeds into a 10 K0 input resistor in series with the base-emitter junction (about 1 kΩ) of the driven transistor. Neglecting the transistor input impedance, this represents a load of about 10 kΩ connected across the output of the driving gate. Further driven gates may be represented by additional 10 K0 resistors in parallel with the first. Each 10 K0 'load' carries approximately one drive unit of current which must come from the supply—and hence flow through the 1 K0 collector load. To maintain our output voltage level above 18 V, this means that the collector current must be limited to 6 mA (i.e., 6 mA × 1 K0 = 6 V), thus restricting fan-out.

If the 10 K0 input resistor is connected in series with a 1 M0 variable resistor, and the base input (drive unit) current measured whilst the collector-emitter voltage is monitored, then the minimum drive unit of current may be determined to saturate the transistor. If each RTL gate uses the increased base input resistance then the fan-out will be increased.

3.12 Diode transistor logic (DTL)

This formed the first really successful range of integrated circuits, and is still used. A simple DTL NAND gate is shown in Fig. 3.11. This simple arrangement is actually made up of a diode AND gate, two level shifting diodes, and an output transistor to provide gain and invert the signal. A

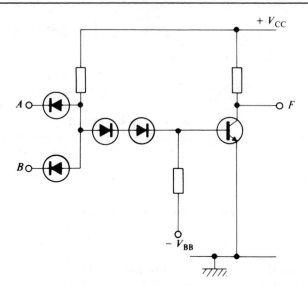

Fig. 3.11. DTL NAND gate.

modification of this basic arrangement which has proved to be particularly successful in IC form is shown in Fig. 3.12, in which an additional transistor is used to reduce power dissipation and increase fan-out.

When the input signals are at logic 1 the input diodes are reverse biased, and current flows through R_1 and R_2 to drive TR 1 into saturation.

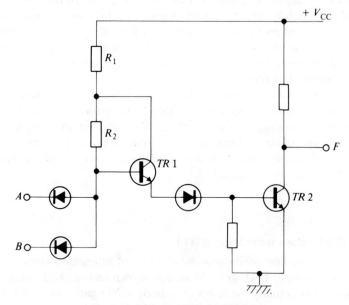

Fig. 3.12. Improved DTL NAND gate.

TR 1 now provides base drive to saturate TR 2, thus the output is at logic 0.

When the input signal at either input is at logic 0, the current is diverted from TR 1 base to switch TR 1 off, which in turn switches TR 2 off, and the output is at logic 1. Therefore, this circuit achieves the NAND logic function.

DTL integrated circuits have typical propagation delays in the region of 25 ns, power dissipations of between 5 and 10 mW per gate, and fan-out of between 8 and 10. Noise immunity is about 1 V. Alternatively, circuits can be made with figures of 50 to 60 ns, 1 to 2 mW per gate, with noise immunity up to 5 V.

Practical Exercise 3C

DISCRETE DTL GATE

Connect up the discrete circuit shown in Fig. 3.13.

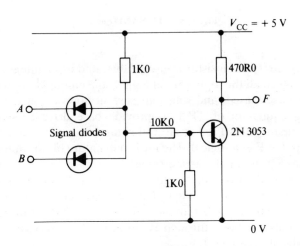

Fig. 3.13. Discrete DTL NAND gate.

Apply all combinations of input signal and determine the truth table for this gate.

Investigate the *quality* of the logic 1 and logic 0.

Investigate the effects of loading the output with other gates.

3.13 Transistor transistor logic (TTL)

This may be considered as a development of DTL, in which the input diodes are replaced by the emitter-base junctions of a multi-emitter transistor as shown in Fig. 3.14.

35

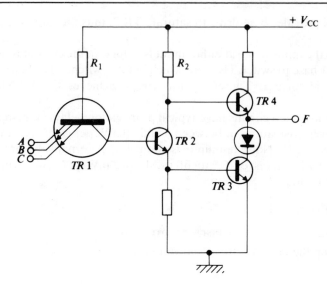

Fig. 3.14. TTL NAND gate.

The multi-emitter transistor is easily fabricated using integrated circuit techniques. When all the inputs are at logic 1, the emitter-base junctions of TR 1 are reverse biased, and sufficient current flows through R_1 and the base-collector junction of TR 1 to provide base drive to switch TR 2 hard-on, so that it holds TR 3 in saturation (and maintains TR 4 at cut-off), and the output at F is at logic 0. The push-pull nature of this output stage is referred to as a 'totem pole' arrangement.

When any or all of the input signals are at logic 0, current flows out of the corresponding emitter of TR 1. This removes the base drive to TR 2, causing it to be cut-off, and removing the drive to TR 3 causing it to be cut-off. Current now flows through R_2 to drive TR 4 into saturation, and the output signal is at logic 1.

TTL has become a particularly popular range, and as a result many variations exist to reduce power dissipation, propagation delays and increase noise margins—this is the range that is particularly discussed in this book. Typical figures for standard TTL gates are 10 mW, 10 ns and 1 V. Some of the variations include:

(a) HTTL (High speed TTL): Typically 10 mW, 6 ns and 1 V respectively.

(b) LTTL (Low power TTL): Typically 1 mW, 35 ns and 1 V respectively.

(c) STTL (Schottky clamped TTL): Typically 20 mW, 3 ns and 0.9 V respectively.

(d) LSTTL (Low power Schottky TTL): Typically 2 mW, 10 ns and 0.8 V respectively.

36

3.14 Sinking and sourcing

When one TTL gate drives another the limiting currents are specified by the manufacturer's. When the output of the driving gate is low (i.e., logic 0), the path for current flow is as shown in Fig. 3.15 (*a*), and the driving gate (TR 3) is said to *sink* the current (provide a path to earth). The specification states that the sinking current is a maximum of 1.6 mA for each driven gate (output). The gates are generally designed to be able to

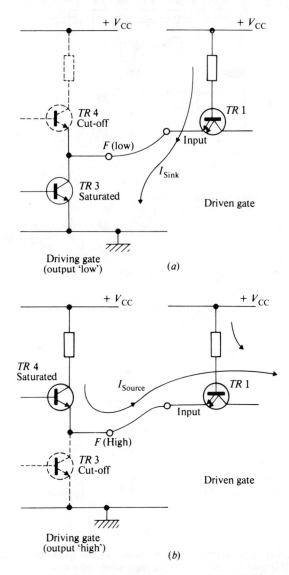

Fig. 3.15. TTL sink and source currents. (*a*) Sinking current.

sink a total of 16 mA, i.e., fan-out 10. When the output of the driving gate is high (i.e., logic 1), the path of the current flow is as shown in Fig. 3.15 (*b*), and the driving gate is said to *source* the current drive for the driven gate (output). The specification states that the source current is a maximum of 40 μA for each load, and the total source current is 400 μA, again giving a fan-out of 10.

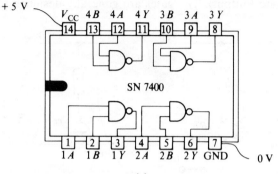

(*a*)

(*b*)

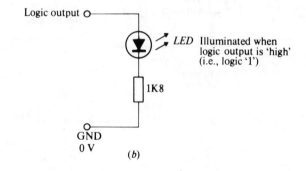

(*c*)

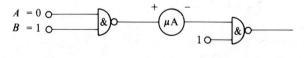

(*d*)

Fig. 3.16. TTL NAND gates (SN 7400). (*a*) **The quadruple 2 input NAND gate IC, (top view),** (*b*) **logic state indicator,** (*c*) **connections to measure 'sink' current,** (*d*) **connections to measure 'source' current.**

Note: The convention that is used in specifications is that current entering the circuit is taken as positive, so that currents leaving the circuit will be negative, e.g., for the TTL gate, when the input is at logic 0. $I_{in} = -1.6$ mA; when the output is at logic 0, $I_{sink} = 16$ mA (total). When the input is at logic 1, $I_{in} = 40$ μA; and when the output is at logic 1, $I_{load} = -400$ μA.

Practical exercise 3D

TTL NAND GATES

Use the SN 7400 IC, Quad 2 i/p NAND gates shown in Fig. 3.16 (*a*). Connect V_{CC} and GND (pins 14 and 7) to +5 V and 0 V respectively.

Apply all combinations of signals to the gate inputs and monitor the output states using LED's as logic state indicators as shown in Fig. 3.16 (*b*). The 1K 8 series resistor will give a reduced illumination, but ensures that the LED does not behave as too much load for the gate outputs.

$$\text{Logic } 1 = +5 \text{ V}$$

$$\text{Logic } 0 = 0 \text{ V}$$

When using NAND gates, all unused inputs must either be connected to used inputs OR must be connected to V_{CC} through a 1K 0 resistor. However, for these simple tests (not permanently connected) no damage should result by connecting unused inputs directly to $+V_{CC}$.

Investigate the quality of the logic 1 and the logic 0 using the digital multimeter (DMM).

Note: It may be necessary to reverse the leads of the DMM to obtain satisfactory voltage readings. The reason for this is that the DMM uses a CMOS circuit, and when making measurements on relatively high impedance circuits spurious voltages are generated about the instrument earth resulting in erroneous readings. Reversal of the leads corrects this, and of course displays a reversal of polarity at the same time.

Measure the sink and source currents when one gate is driving another gate in the 1 and 0 states by connecting the arrangements shown in Fig. 3.16 (*c*) and (*d*), and check that these currents are less than 1.6 mA and 40 μA respectively.

3.15 TTL recognition—device numbering

Each part of a TTL device symbolisation can be divided into distinctly separate parts, each of which tell us something about the device.

Example

SN   74 H 107 N

SN   Semiconductor Network.

39

74 TTL is manufactured to meet THREE temperature ranges; the most widely used being:

Series 54 −55°C to +125°C   military

Series 74   0°C to +70°C   industrial.

Supply voltage tolerances are also different:

Series 54   4.5 to 5.5 V

Series 74   4.75 V to 5.25 V

H   High speed device. Variations of this include:

   L   Low power

   S   Schottky

   LS   Low power Schottky

   NO LETTER   Standard TTL

107 Device Function (TWO or THREE numbers).

   107   Dual J-K flip-flop.

N   Package Type.

   N 14, 16, 24 pin DIL plastic (MOST COMMON).

3.16 Specification—TTL data sheet

A typical data sheet for the basic TTL Quad 2 i/p NAND gate is shown in Fig. 3.17. These data sheets are usually prepared to cover *both* the series 54 *and* series 74 devices—the main differences between which we have already noted in para. 3.15.

The pin-out diagrams for flat package and dual in line package are included (as viewed from the top of the device). Note that the pin numbers for the connections are different between the two packages.

Data sheets always include the supply voltages and temperature ranges, together with the fan out. This is followed by the worst case parameters—these values we have already discussed in para. 3.4 and para. 3.14.

Finally the switching characteristics are given as the propagation delay time *low to high* logic level and *high to low* logic level with a standard load of 400 Ω resistance and 15 pF capacitance.

3.17 Emitter coupled logic (ECL)

This logic family differs in principle from previous types in that the transistors are not all operated in the saturated mode. Due to the increased speed of operation this is quite widely used in fast computing circuits. A typical ECL gate is shown in Fig. 3.18, which is capable of performing both the OR and the NOR logic functions.

When both inputs A and B are low, TR 1 and TR 2 are cut off and their collector voltage is high. TR 3 is switched on by the bias voltage V_{REF} applied to its base, and its collector voltage is low. The high collector voltages of TR 1 and TR 2 causes TR 4 to switch on, which causes output

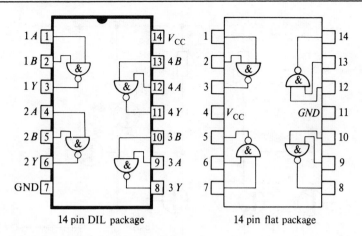

14 pin DIL package   14 pin flat package

Absolute max. supply voltage V_{CC} = 7 V
Absolute max. input voltage V_{IN} = 5.5 V
Normal operating supply voltage 54 series V_{CC} = 5 V $\pm$ 0.5 V
Normal operating supply voltage 74 series V_{CC} = 5 V $\pm$ 0.25 V
Operating temperature 54 series, T_A = $-$55 °C to + 125 °C
Operating temperature 74 series, T_A = 0 °C to + 70 °C

Electrical characteristics:

Input voltage: Low level , V_{IL} = 0.8 V max.
                High level , V_{IH} = 2.0 V min.
Output voltage: Low level , V_{OL} = 0.4 V max. (typ. 0.22 V)
                High level , V_{OH} = 2.4 V min. (typ. 3.3 V)
Input current : Low level , (V_{CC} = max. V_{IN} = 0.4 V) I_{IL} = $-$ 1.6 mA
                High level , (V_{CC} = max. V_{IN} = 2.4 V) I_{IH} = 40 μA

Switching characteristics:

Load R_L = 400 Ω , C_L = 15 pF
Propagation delay time (low to high logic level) t_{PLH} = 22 ns max.(typ. 11 ns)
Propagation delay time (high to low logic level) t_{PHL} = 15 ns max.(typ. 7 ns)

Fig. 3.17. Data sheet for TTL circuit type SN 7400.

F_1 to go high. At the same time, the low collector voltage of TR 3 is insufficient to switch on TR 5, and the output F_2 is low.

When any input is high, the emitter-base voltage of TR 3 is reduced, which causes TR 3 to be cut off, and the collector voltage of TR 3 becomes greater than the collector voltages of TR 1 and TR 2. Under these conditions the output F_2 is high and the output at F_1 is low. Therefore, the OR function is obtained at F_2, and the NOR function at F_1.

Typical propagation delays of 2 ns are available, with power dissipation of 25 mW. High fan-out of about 30, but noise immunity is low at about 0.2 V. Higher speed circuits are available at about 1 ns.

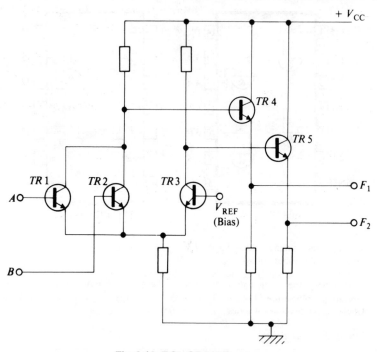

Fig. 3.18. ECL OR/NOR gate.

3.18   MOS logic

The main advantages of MOS techniques are small area of fabrication, and low power dissipation. These advantages lend themselves to large scale integration (LSI) such as computer memories. A simple MOS NOT gate is shown in Fig. 3.19 using p channel FET's. FET 2 behaves as a load resistance for FET 1. When the input signal is low (negative for p channel

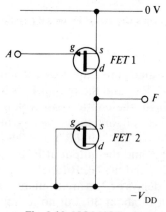

Fig. 3.19. MOS NOT gate.

FET's), FET 1 will be saturated and the output F will be high (0V). When the input is high, FET 1 is cut off and the output F is low (i.e., $-V_{DD}$), thus producing inversion.

A simple MOS NAND gate is shown in Fig. 3.20, which also uses p channel devices. When either of the inputs A or B is low (i.e., negative), that FET is saturated, and the output F is high. When both inputs at A and B are high, FET 1 and FET 2 are cut off and the output F is low (i.e., $-V_{DD}$), thus producing the NAND logic function.

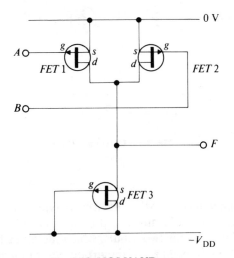

Fig. 3.20. MOS NAND gate.

3.19 CMOS logic

This logic family is based on both n channel and p channel MOS devices used in a *complementary* symmetrical arrangement. A simple CMOS NOR gate is shown in Fig. 3.21, which uses positive logic.

When both input signals A and B are low, FET 3 and FET 4 are cut off and FET 1 and FET 2 are switched on, giving a 'high' output at F. When the input signal at A is low, and the input signal at B is high, FET 1 and FET 3 are saturated and FET 2 and FET 4 are cut off, giving a low output at F. This is also the case when A is high and B is low. When both input signals at A and B are high, FET 1 and FET 2 are cut off, and FET 3 and 4 are saturated, again giving a low output at F.

The quality of the high and low states are very nearly $+V_{DD}$ and very nearly 0V respectively. This means that CMOS circuits have a very high noise immunity—typically 20% of the supply voltage. This family can operate with supply voltages ranging between 3 V and 15 V, but for applications such as electronic watch circuits the supply requirements are only about 5 μA at between 1 V and 1.5 V, and special care must be taken to reduce the threshold voltage levels.

43

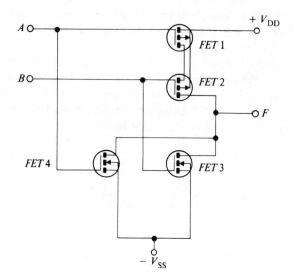

Fig. 3.21. CMOS NOR gate.

The input impedance of MOS devices is typically 10^{12} ohms with a typical capacitance of 5 pF, and the propagation delay depends on the fan-out—typically 20 ns for two inputs, which increases by about 5 ns for each 5 pF load. If high speed is not important, fan-out may be increased to about 50.

When handling CMOS devices great care must be taken since they are very susceptible to damage from excessive voltage caused by static electricity and equipment which is not correctly earthed. These devices are often supplied with their pins embedded in conductive foam, KEEP THIS INTACT UNTIL THE CIRCUIT IS TO BE CONNECTED. *DO NOT USE PLASTIC OR NYLON MATERIAL* since these generate high static voltages. Similarly, take care with clothing—many of the man-made fibres used today also generate large amounts of static electricity. In addition, all equipment used must be properly earthed. Working with CMOS has therefore created new problems, most of which can be overcome without too much difficulty. One method which has been used very successfully is that of a copper plate working surface properly earthed together with all other equipment—even to the extent of conductive wrist-bands connected to the same earth. However, most CMOS IC's available today have some internal protection (buffered inputs), but, even with these it is important to connect the supply pins first. Finally, DO NOT LEAVE UNUSED INPUTS FLOATING, always connect unused inputs either to used inputs, or to a supply line through a 220 K0 resistor. Input signals must not be applied until the power supply is connected, and is ON.

Some of the disadvantages of CMOS have been emphasised here in an attempt to stress the importance of *handle with care*, since so many heartaches have been caused by not observing simple precautions.

The main advantages of CMOS are packing density, wide supply voltage range, high noise immunity, and low power consumption. Operating from a 5 V supply the propagation delay is 35 ns, power dissipation 10 nW and noise immunity 2 V. These factors, together with reducing costs has led to the wide application of CMOS IC's for both digital and linear systems. However, circuits and applications discussed in this book apply to TTL (Bipolar) 74 series digital IC's, although the *logic* applies to CMOS, and any other system.

4 Logic networks

4.1 NAND and NOR gates

In practice, the particular range of integrated circuits contain only a limited range of gate types. Generally, the restrictions are that only NAND gates, or only NOR gates are used.

Although a particular system may contain all NAND gates only, or all NOR gates only, it must be remembered that the *basic* logic functions needed to be performed are: AND, OR and NOT. These basic logic functions may be performed using NAND gates only as shown in Fig. 4.1 (*a*) and (*b*).

The basic logic functions may also be performed by using NOR gates as shown in Fig. 4.2 (*a*) and (*b*).

Logic networks may then be constructed using these arrangements to produce the basic logic functions. An important step is now necessary—to

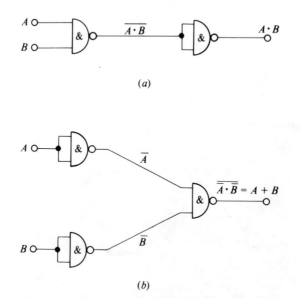

(*a*)

(*b*)

Fig. 4.1. Basic logic functions using NAND gates only.
(*a*) AND function using NAND gates only, (*b*) OR function using NAND gates only.

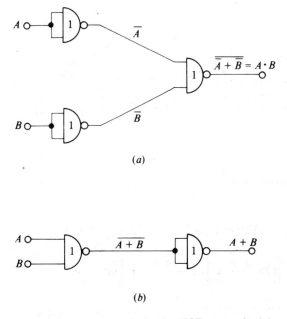

(a)

(b)

**Fig. 4.2. Basic logic functions using NOR gates only. (*a*)
AND function using NOR gates only, (*b*) OR function
using NOR gates only.**

check the resultant network to eliminate redundant gates, i.e., *minimization*.

Example 4.1

Construct a logic network using NAND gates only for the logic function F.

$$F = A.B + C.D$$

Considering the basic logic gates, the network is as shown in Fig. 4.3 (*a*).

If each separate gate is now replaced by its NAND gate equivalent (as shown in Fig. 4.1), then the network is as shown in Fig. 4.3 (*b*).

Now, close examination of the logic network shown in Fig. 4.3 (*b*) reveals the existence of redundant gates, so that the equivalent logic network is as shown in Fig. 4.3 (*c*) using NAND gates only.

Example 4.2

Construct a logic network using NOR gates only for the logic function F.

$$F = (A + B).(C + D)$$

Firstly, consider the network using basic logic gates to represent the logic function F as shown in Fig. 4.4 (*a*).

47

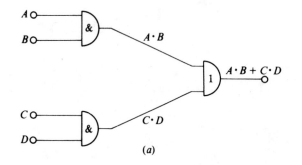

(a)

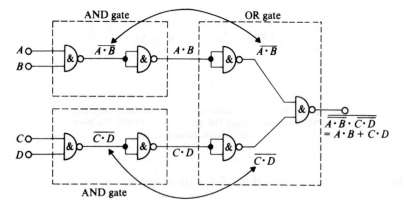

(b)

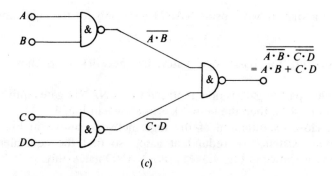

(c)

Fig. 4.3. NAND gate implementation, and minimization. (a) Using basic logic gates, (b) using NAND gates only, (c) resultant NAND gate only network.

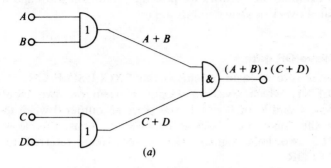

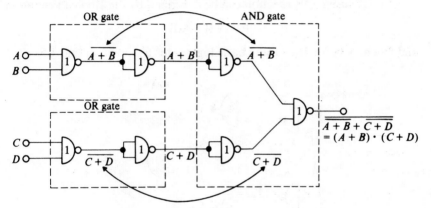

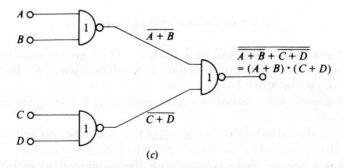

Fig. 4.4. NOR gate implementation, and minimization. (a) using basic logic gates, (b) replacing basic gates by NOR gate equivalents. (c) minimal NOR gate only logic network.

Secondly, replace each basic logic gate by its NOR equivalent (as shown in Fig. 4.2) and the logic network becomes as shown in Fig. 4.4 (*b*).

Finally, examine the network for possible redundant gates and redraw the minimal network as shown in Fig. 4.4 (*c*).

4.2 Exclusive OR gate

A very useful gate in logic systems is the EXCLUSIVE OR, or NOT EQUIVALENT, which gives an output 1 when the two inputs are different, i.e., 1 and 0, or 0 and 1, and gives an output 0 when the two inputs are the same, i.e., both 0 or both 1. Thus, this logic gate COMPARES two logic signals, and is sometimes referred to as a COMPARATOR.

The Exclusive OR function can be described by the Boolean equation:

$$F = A.\bar{B} + \bar{A}.B$$

and this may be realised using basic logic gates as shown in Fig. 4.5.

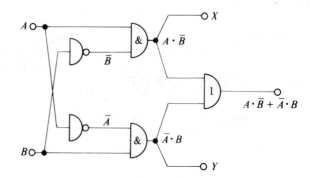

Fig. 4.5. Basic Exclusive OR logic network.

The signal output at X will be 1 when A > B (A greater than B), i.e., when A = 1, B = 0, and the signal output at Y will be 1 when A < B (A less than B), i.e., when A = 0, B = 1.

The Exclusive OR function can be realised using NAND gates only, as shown in Fig. 4.6.

You are advised to follow through all of the logic networks and confirm that the Boolean algebra has been correctly applied in each case, in this way you will become more familiar with the manipulations of Boolean algebra.

Practical exercise 4A

EXCLUSIVE OR

Connect up the exclusive OR logic networks shown in Fig. 4.6 using 2 i/p NAND gates, i.e., SN 7400, Quad 2 i/p NAND.

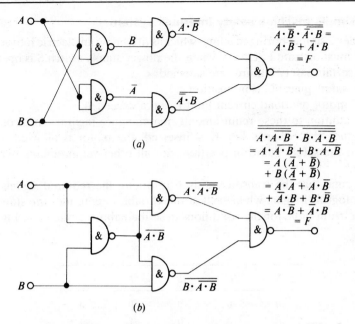

$$\overline{A \cdot \overline{B} \cdot \overline{\overline{A} \cdot B}} =$$
$$A \cdot \overline{B} + \overline{A} \cdot B$$
$$= F$$

$$\overline{\overline{A \cdot \overline{A \cdot B}} \cdot \overline{B \cdot \overline{A \cdot B}}}$$
$$= A \cdot \overline{A \cdot B} + B \cdot \overline{A \cdot B}$$
$$= A \, (\overline{A} + \overline{B})$$
$$+ \, B \, (\overline{A} + \overline{B})$$
$$= A \cdot \overline{A} + A \cdot \overline{B}$$
$$+ \, \overline{A} \cdot B + B \cdot \overline{B}$$
$$= A \cdot \overline{B} + \overline{A} \cdot B$$
$$= F$$

Fig. 4.6. Exclusive OR networks using NAND gates only. (*a*) Using 5 NAND gates, (*b*) using 4 NAND gates.

Apply all combinations of input signals to A and B and observe the logic state at the output using LED logic state indicators. Hence, construct the truth table for this logic function.

Practical exercise 4B

COMPARATOR

Connect up the comparator network shown in Fig. 4.7 using $\frac{3}{4} \times$ SN 7400, Quad 2 i/p NAND and $\frac{1}{3} \times$ SN 7404, Hex. inverter, and verify the logical states of the outputs (using the LED logic state indicators) for all combinations of input signals.

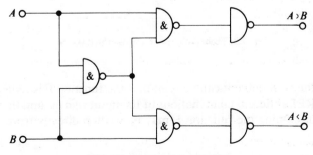

Fig. 4.7. Comparator network.

51

4.3 Simple machine safety interlock system

Consider a simple boring machine which is driven by an electric motor. The motor must operate F, ONLY when the power supply switch S is operated AND certain safety features are satisfied:

(a) safety guard G is in position,

(b) motor overload current limit L is not exceeded.

In addition to these requirements, maintenance facilities must be provided such that when a key K is inserted, the motor is allowed to run without the safety guard in position, but all other requirements being as before.

Assume that the transducers used to obtain the required signals produce a logical 1 signal when each of the variables are in the *safe* state.

The truth table for the conditions described above is shown in Fig. 4.8.

S	G	L	K	F
0	0	0	0	0
0	0	0	1	0
0	0	1	0	0
0	0	1	1	0
0	1	0	0	0
0	1	0	1	0
0	1	1	0	0
0	1	1	1	0
1	0	0	0	0
1	0	0	1	0
1	0	1	0	0
1	0	1	1	1
1	1	0	0	0
1	1	0	1	0
1	1	1	0	1
1	1	1	1	1

Fig. 4.8. Truth table for safety interlock system.

From the truth table it can be seen that the function F is realised (logic 1) for THREE different combinations of the input signals, and the Boolean equation describing the function F may be written directly from the truth table:

$$F = S.\bar{G}.L.K + S.G.L.\bar{K} + S.G.L.K$$

52

This Boolean equation describes the function in terms of the basic logic operations, for which the logic network is as shown in Fig. 4.9. This network would appear to use a rather excessive number of gates.

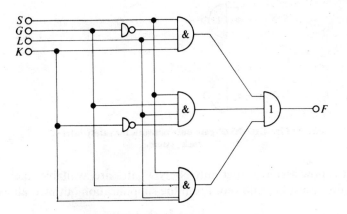

Fig. 4.9. Safety interlock using basic gates.

If it is assumed that only NOR gates are available, then Boolean algebra techniques may be applied to the above Boolean equation in order to simplify it, and to rearrange the equation into a form which can be directly implemented by NOR gates:

$$F = S.\bar{G}.L.K + S.G.L.\bar{K} + S.G.L.K$$

$$\therefore \quad F = S.\bar{G}.L.K + S.G.L (\bar{K} + K)$$

$$\therefore \quad F = S.\bar{G}.L.K + S.G.L$$

$$\therefore \quad F = S.L. (\bar{G}.K + G)$$

$$\therefore \quad F = S.L. (K + G) \quad \text{since } (\bar{G}.K + G) = (\bar{G} + G).(K + G) = (K + G)$$

This is a simplified expression using basic logic functions.

$$\bar{F} = \overline{S.L. (K + G)}$$

$$\bar{F} = \bar{S} + \bar{L} + \overline{K + G}$$

and, finally complementing to reproduce the original function F, gives:

$$F = \overline{\bar{S} + \bar{L} + \overline{K + G}}$$

which is an expression which may be directly implemented using NOR gates only as shown in Fig. 4.10.

53

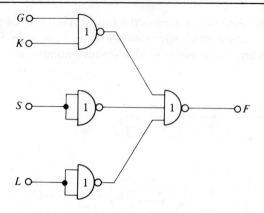

Fig. 4.10. NOR gate only network for safety interlock system.

If it is now assumed that only NAND gates are available, then similar procedures must be followed. From the simplification outlined above:

$$F = S.L. (K + G)$$

$$\therefore \quad F = S.L. \overline{\overline{K}.\overline{G}}.$$

$$\therefore \quad \overline{F} = \overline{S.L. \overline{\overline{K}.\overline{G}}}.$$

The right hand side of this equation can now be directly implemented by NAND gates only. But, this produces the complement of the desired function, so that it is necessary to use an additional inverter to reproduce the original function F as shown in Fig. 4.11.

Note: The techniques of deriving minimal NAND or minimal NOR networks using Boolean algebra can often be somewhat obscure and laborious, and although very useful it requires a reasonable understanding of the laws of Boolean algebra. Chapter 5 deals with a very widely used method of achieving similar objectives—again, an understanding of the basic principles involved is essential.

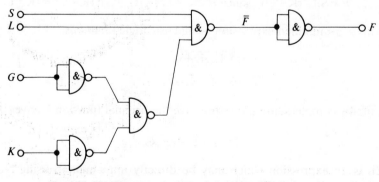

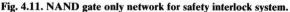

Fig. 4.11. NAND gate only network for safety interlock system.

Practical exercise 4C

SAFETY INTERLOCK SYSTEM

Connect up the logic network shown in Fig. 4.10 using the SN 7402, Quad 2 i/p NOR gates. Apply all the combinations of input signals and verify the truth table shown in Fig. 4.8.

Connect up the logic network shown in Fig. 4.11 using $1 \times$ SN 7400, Quad 2 i/p NAND gate and $\frac{1}{2} \times$ SN 7420, Dual 4 i/p NAND gate.

Apply all the combinations of the input signals and verify the truth table shown in Fig. 4.8.

5 Karnaugh maps

5.1 Introduction

This is a graphical method of representing the truth table of a given logical function. The Karnaugh map is a rectangular diagram, the area of which is divided into *cells*, where:

$$\text{Total number of cells in the map} = 2^N$$
$$\text{where } N = \text{number of logical variables}$$

Each logical variable in a Karnaugh map is represented by *half* the total area, and *its complement by the other half.*

Karnaugh maps for one, two and three variable systems are as shown in Fig. 5.1 (*a*), (*b*) and (*c*) respectively, and the Boolean expression is written *inside* each cell—derived in a similar way to the game Battleships.

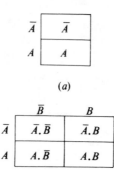

(*a*)

(*b*)

(*c*)

Fig. 5.1. Karnaugh maps for one, two and three variables. (*a*) One variable 2^1 (2^1 cells = 2), (*b*) two variables (2^2 cells = 4), (*c*) three variables, (2^3 cells = 8).

The Karnaugh map for a four variable system is shown in Fig. 5.2 (a). Again, the Boolean expression for each combination of the four variables is included inside each cell.

If we now use a logical 1 to represent each of the variables A, B, C and D, and logical 0 to represent the complements, i.e., $\bar{A}$, $\bar{B}$, $\bar{C}$ and $\bar{D}$ and redraw the Karnaugh map to that shown in Fig. 5.2 (b), the resulting diagram should strictly be called a *Veitch diagram*. However, to avoid the possibility of confusion, we still tend to refer to all such diagrams as Karnaugh maps.

	$\bar{C}$	$\bar{C}$	C	C	
$\bar{A}$	$\bar{A}\bar{B}\bar{C}\bar{D}$	$\bar{A}\bar{B}\bar{C}D$	$\bar{A}\bar{B}CD$	$\bar{A}\bar{B}C\bar{D}$	$\bar{B}$
$\bar{A}$	$\bar{A}B\bar{C}\bar{D}$	$\bar{A}B\bar{C}D$	$\bar{A}BCD$	$\bar{A}BC\bar{D}$	B
A	$AB\bar{C}\bar{D}$	$AB\bar{C}D$	$ABCD$	$ABC\bar{D}$	B
A	$A\bar{B}\bar{C}\bar{D}$	$A\bar{B}\bar{C}D$	$A\bar{B}CD$	$A\bar{B}C\bar{D}$	$\bar{B}$
	$\bar{D}$	D	D	$\bar{D}$	

(a)

AB \ CD	00	01	11	10
00	0000	0001	0011	0010
01	0100	0101	0111	0110
11	1100	1101	1111	1110
10	1000	1001	1011	1010

(b)

Fig. 5.2. Karnaugh map and Veitch diagram for four variables. (a) Karnaugh map, (b) Veitch diagram.

In Fig. 5.2 (b), the cells within the map have been completed using the same techniques as before, except that this time the logical levels of 0 and 1 have been used—producing an array which appears to be similar to a 4-bit binary number. The reader is advised to confirm that the two diagrams shown in Fig. 5.2 are in fact different methods of representing the same truth table.

If the positional 'weights' of the digits in the cells in Fig. 5.2 (b) are ignored, then it should be observed that *adjacent cells* in a *Karnaugh map differ by only one binary digit*. In this sense, cells at the top are adjacent to cells at the bottom, and cells on the left are adjacent to cells on the right. Confirm that this is correct before proceeding.

5.2 Function mapping

Mapping is a graphical method of representing logical equations. This technique is widely used to prove Boolean theorems, to design logical networks and to assist in the minimisation of logic gates in certain networks.

5.3 Cell looping

This very important principle is best described using an example:

Example 5.1

Suppose that a particular logic function is described by the Boolean equation:

$$F = A.B.C.D + \bar{A}.B.C.D + \bar{A}.B.\bar{C}.D + \bar{A}.\bar{B}.\bar{C}.D$$

The truth table for this equation is therefore as shown in Fig. 5.3, in which F (the function) is shown as a logical 1 for each of the conditions described by each of the terms in the Boolean equation above, and logical 0 for all other conditions.

The Karnaugh map may be drawn for this system—either directly from the original Boolean equation, or from the truth table. *BUT*, note that we now only write a logical 1 in each cell which describes *one* four variable

A	B	C	D	F
0	0	0	0	0
0	0	0	1	1
0	0	1	0	0
0	0	1	1	0
0	1	0	0	0
0	1	0	1	1
0	1	1	0	0
0	1	1	1	1
1	0	0	0	0
1	0	0	1	0
1	0	1	0	0
1	0	1	1	0
1	1	0	0	0
1	1	0	1	0
1	1	1	0	0
1	1	1	1	1

Fig. 5.3. Truth table for Example 5.1.

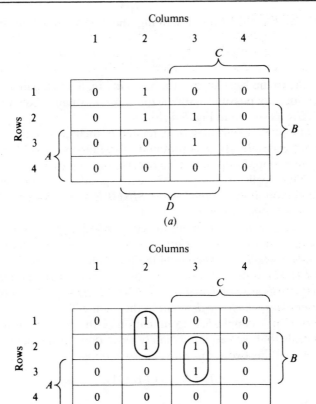

Fig. 5.4. Karnaugh maps for Example 5.1.

term in the equation, and logical 0 in all other cells, as shown in Fig. 5.4 (*a*).

Now, consider the *adjacent cells* in rows 1 and 2 of column 2. These two cells correspond to the terms $\bar{A}.B.\bar{C}.D$ and $\bar{A}.\bar{B}.\bar{C}.D$ in the above Boolean equation. These two terms may be simplified using Boolean algebra techniques:

$$\bar{A}.B.\bar{C}.D + \bar{A}.\bar{B}.\bar{C}.D = \bar{A}.\bar{C}.D.(B + \bar{B})$$

$$= \bar{A}.\bar{C}.D$$

Similarly, consider the adjacent cells in rows 2 and 3 of column 3, which correspond to the terms $A.B.C.D$ and $\bar{A}.B.C.D$. These two terms may also be simplified:

$$A.B.C.D + \bar{A}.B.C.D = B.C.D.(A + \bar{A})$$

$$= B.C.D$$

59

Therefore, the equation describing the logical function F may be rewritten as:

$$F = \bar{A}.\bar{C}.D + B.C.D$$

Returning to the Karnaugh map shown in Fig. 5.4 (*a*), pairs of adjacent cells which identify possible simplifications are normally shown *LOOPED TOGETHER* as shown in Fig. 5.4 (*b*).

Note: This looping of cells may be extended to include 4 cells, 8 cells, etc., and we generally try to accommodate the largest possible number of cells in a loop. Loops may also overlap one another.

When *TWO* cells are looped, the corresponding *TWO* terms in the Boolean equation are *combined into a SINGLE term—which excludes the variable which is different* in the two cells.

When *FOUR* cells are looped, the corresponding *FOUR* terms in the Boolean equation are *combined into a SINGLE term—which excludes the TWO variables which are different* in the four cells.

When *EIGHT* cells are looped, the corresponding *EIGHT* terms in the Boolean equation are *combined into a SINGLE term—which excludes the THREE variables which are different* in the eight cells.

Therefore, in Example 5.1, examination of the Karnaugh map in Fig. 5.4 (*b*), reveals that we can immediately write down a simplified equation for F which contains two terms only, i.e. two loops.

$$F = \bar{A}.\bar{C}.D + B.C.D.$$

Example 5.2

Draw the Karnaugh map for the following Boolean equation:

$$F = A.\bar{B}.\bar{C} + A.B.\bar{C}$$

Use cell looping techniques to simplify this function.
The Karnaugh map is shown in Fig. 5.5.
From the *one* loop shown:

$$F = A.\bar{C}$$

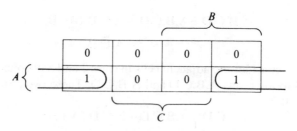

Fig. 5.5. Karnaugh map for Example 5.2.

Example 5.3

Draw the Karnaugh map for the function F described by the Boolean equation:

$$F = A.\bar{B}.\bar{C}.D + A.\bar{B}.\bar{C}.\bar{D} + \bar{A}.B.C.D + \bar{A}.B.\bar{C}.D$$

Use cell looping techniques to simplify this function.
The Karnaugh map is shown in Fig. 5.6.

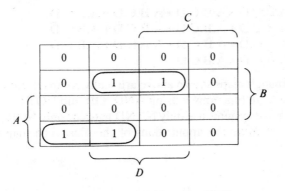

Fig. 5.6. Karnaugh map for Example 5.3.

From the *two* loops shown:

$$F = A.\bar{B}.\bar{C} + \bar{A}.B.D.$$

Example 5.4

Draw the Karnaugh map for the function F described by the Boolean equation:

$$F = A.\bar{B}.\bar{C}.\bar{D} + A.\bar{B}.C.\bar{D} + \bar{A}.\bar{B}.C.\bar{D} + \bar{A}.\bar{B}.\bar{C}.\bar{D}.$$

Use cell looping techniques to simplify this function.
The Karnaugh map is shown in Fig. 5.7.

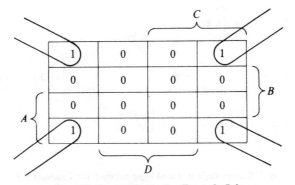

Fig. 5.7. Karnaugh map for Example 5.4.

61

From the *one* loop shown:

$$F = \bar{B}.\bar{D}.$$

Exercises

Plot the Karnaugh maps for the functions described by the following Boolean equations and use cell looping techniques to simplify these functions:

1. $F = A.\bar{B}.C.D + A.B.C.D + A.\bar{B}.\bar{C}.D + A.B.\bar{C}.D.$
2. $F = A.B.\bar{C}.\bar{D} + \bar{A}.B.\bar{C}.\bar{D} + \bar{A}.B.C.\bar{D} + A.B.C.\bar{D}.$
3. $F = \bar{A}.\bar{B}.\bar{C}.\bar{D} + \bar{A}.\bar{B}.\bar{C}.D + A.\bar{B}.\bar{C}.\bar{D} + A.\bar{B}.\bar{C}.D$
4. $F = \bar{A}.\bar{B}.C.\bar{D} + \bar{A}.\bar{B}.C.D$

The Karnaugh maps shown above are really compact truth tables which describe the logic functions F using AND, OR and NOT logic operations. In practice, most systems use only NAND gates or only NOR gates, so that we need to examine the applications of the Karnaugh map to meet this requirement.

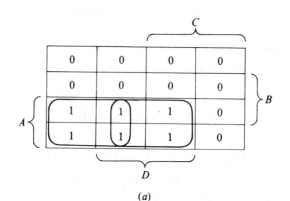

(a)

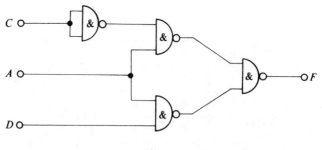

(b)

Fig. 5.8. Karnaugh map and logic network for Example 5.5.
(a) Karnaugh map, (b) NAND gate only network.

Example 5.5

Suppose that we are given the Karnaugh map shown in Fig. 5.8 (*a*) describing the logical function F.

Assume that we require a logic network comprising NAND gates only to perform this logical function.

From the *two* loops shown, the Boolean equation describing this function is:

$$F = A.D + A.\bar{C}$$

which describes the function using AND, OR and NOT operations.

Now, using Boolean algebra techniques:

$$\bar{F} = \overline{A.D + A.\bar{C}}$$
$$= \overline{\overline{A.D}.\overline{A.\bar{C}}}$$
$$\therefore \quad F = \overline{\overline{A.D}.\overline{A.\bar{C}}}$$

This equation now contains terms which may be directly implemented using NAND gates only, as shown in Fig. 5.8 (*a*).

Example 5.6

Consider the same logical function as in Example 5.5. The Karnaugh map is redrawn as shown in Fig. 5.9 (*a*).

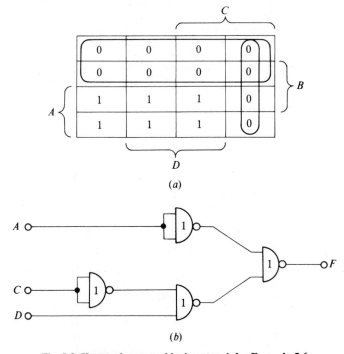

(*a*)

(*b*)

Fig. 5.9. Karnaugh map and logic network for Example 5.6.
(*a*) Karnaugh map, (*b*) NOR gate only network.

Assume that we require a logic network comprising NOR gates only to perform this logical function.

The Boolean expression which describes the function F is more conveniently derived by looping cells containing 0's when we require a NOR gate only network.

Thus, from the *two* loops shown in Fig. 5.9 (*a*), the Boolean equation describing the *complement* of the function F is:

$$\bar{F} = \bar{A} + C.\bar{D}$$

which could be implemented using the basic AND, OR and NOT gates.

This Boolean equation may be rewritten:

$$\bar{F} = \bar{A} + \overline{\bar{C} + D}$$

and, complementing to produce the original function F gives:

$$F = \overline{\bar{A} + \overline{\bar{C} + D}}$$

This equation now contains terms which may be directly implemented using NOR gates only as shown in Fig. 5.9 (*b*).

Some situations occur in practice where a cell (or group of cells) may contain a logical 0 *OR* a logical 1. In these cells we write an asterisk '*' so

X		Y		F
A	B	C	D	
0	0	0	0	0
0	0	0	1	0
0	0	1	0	0
0	0	1	1	0
0	1	0	0	1
0	1	0	1	0
0	1	1	0	0
0	1	1	1	0
1	0	0	0	1
1	0	0	1	1
1	0	1	0	0
1	0	1	1	0
1	1	0	0	1
1	1	0	1	1
1	1	1	0	1
1	1	1	1	0

Fig. 5.10. Truth table for Example 5.7.

that when we examine the map for possible loops, we can include or exclude these cells as required to complete the loops.

Example 5.7

Consider a logic system which has FOUR signal inputs designated A, B, C and D. It may be assumed that the logic levels at A and B represent a binary number X, in which A is the most significant digit (MSD). Similarly the logic levels at C and D represent a second binary number Y, in which C is the MSD.

The conditions of this system for which the output F is logical 1 are satisfied when the binary number X is greater than the binary number Y. The output F is logical 0 for all the remaining values of X and Y.

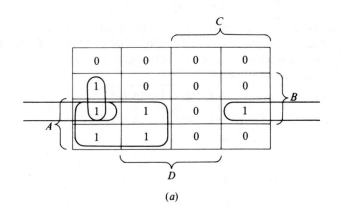

(a)

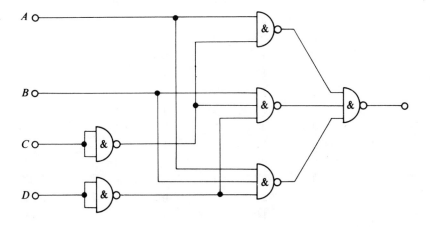

(b)

Fig. 5.11. Karnaugh map and logic network for Example 5.7. (a) Karnaugh map, (b) NAND gate only logic netowrk.

65

(a) Draw the Karnaugh map for this logic system, and use cell looping techniques to derive a simple Boolean equation to describe the function F.

(b) Derive a logic network using NAND gates only which is capable of performing this function.

(a) It is advisable, in this case, to draw up the truth table first, as shown in Fig. 5.10.

The Karnaugh map may now easily be constructed from the Truth Table as shown in Fig. 5.11 (a).

Since we are ultimately required to derive a NAND gate network, consider the cells containing 1's.

From the *THREE* loops shown in the Karnaugh map, the Boolean equation is:

$$F = A.\bar{C} + B.\bar{C}.\bar{D} + A.B.\bar{D}$$

(b) The above equation describes the function F in terms of the *basic* logic operations, i.e., AND, OR and NOT. To derive the equation which describes this function in terms of NAND gates, proceed as follows:

$$\bar{F} = \overline{A.\bar{C} + B.\bar{C}.\bar{D} + A.B.\bar{D}}$$

$$= \overline{A.\bar{C}} \cdot \overline{B.\bar{C}.\bar{D}} \cdot \overline{A.B.\bar{D}}$$

X			Y	F
A	B	C	D	
0	0	0	0	1
0	0	0	1	0
0	0	1	0	0
0	0	1	1	0
0	1	0	0	1
0	1	0	1	1
0	1	1	0	0
0	1	1	1	0
1	0	0	0	1
1	0	0	1	1
1	0	1	0	1
1	0	1	1	0
1	1	0	0	1
1	1	0	1	1
1	1	1	0	1
1	1	1	1	1

Fig. 5.12. Truth table for Example 5.8.

Finally, complementing to reproduce the original function F gives:

$$F = \overline{\overline{A.\bar{C}} . \overline{B.\bar{C}.\bar{D}} . \overline{A.B.\bar{D}}}$$

and the NAND gate only network may now be drawn as shown in Fig. 5.11 (*b*).

Example 5.8

Consider a logical system which is similar to that described in Example 5.7, in which the output function F is logical 1 when the binary number X is greater than, or equal to, the binary number Y. The output F is logical 0 for all the remaining values of X and Y.

Derive the new Boolean equation, and a logic network using NOR gates only.

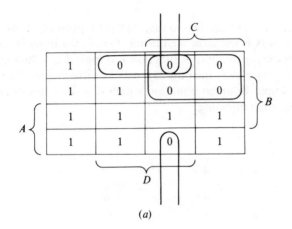

(*a*)

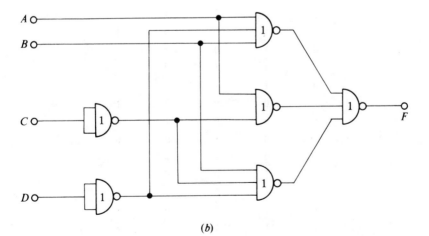

(*b*)

Fig. 5.13. Karnaugh map and logic network for Example 5.8. (*a*) Karnaugh map, (*b*) NOR gate only network.

The truth table is shown in Fig. 5.12, from which the Karnaugh map is easily drawn as shown in Fig. 5.13 (a).

Since a NOR gate only network is required, we loop cells containing 0's to give:

$$\bar{F} = \bar{A}.\bar{B}.D + \bar{A}.C + \bar{B}.C.D$$

$$\therefore \quad F = \overline{\bar{A}.\bar{B}.D + \bar{A}.C + \bar{B}.C.D}$$

This equation may now be converted into a suitable form to be directly implemented using NOR gates:

$$F = \overline{\overline{A + B + \bar{D}} + \overline{A + \bar{C}} + \overline{B + \bar{C} + \bar{D}}}$$

and the logic network is as shown in Fig. 5.13 (b).

Example 5.9

The four inputs A, B, C and D to a logic system represent a four-bit binary number in which A is the most significant digit. If the input is less than or equal to denary eight the output function F is logical 1. When the input is greater than denary eleven the output F may be either logical 1 or logical 0.

Derive a simple Boolean equation which describes this operation, by the use of Karnaugh mapping.

A	B	C	D	F
0	0	0	0	1
0	0	0	1	1
0	0	1	0	1
0	0	1	1	1
0	1	0	0	1
0	1	0	1	1
0	1	1	0	1
0	1	1	1	1
1	0	0	0	1
1	0	0	1	0
1	0	1	0	0
1	0	1	1	0
1	1	0	0	*
1	1	0	1	*
1	1	1	0	*
1	1	1	1	*

Fig. 5.14. Truth table for Example 5.9.

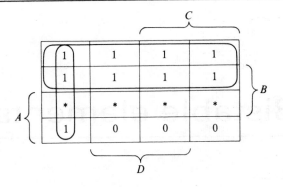

Fig. 5.15. Karnaugh map for Example 5.9.

The truth table is drawn up as shown in Fig. 5.14. From the truth table the Karnaugh map may be easily drawn as shown in Fig. 5.15.

From the *TWO* loops shown in the Karnaugh map:

$$F = \bar{A} + \bar{C}.\bar{D}$$

As an exercise in the techniques described, you may now try to derive logic networks using NAND gates only and NOR gates only for the system described in Example 5.9.

Note: It is possible to construct the Karnaugh maps directly from the descriptions of the logic systems in the Examples 5.7 to 5.9, the truth tables have only been constructed to simplify the construction of the Karnaugh maps.

6 Bistable elements

6.1 Introduction

Bistable elements have *TWO* stable operating states, in which the application of a signal causes the device to change from one stable operating state to the other. Devices of this type are simple memories, since the state of the output at any instant can be used to deduce the state of the input at a previous instant. These devices may also be referred to as *flip-flops*, *static memories*, etc., and systems which use them are known as *sequential logic systems*.

6.2 The S–R flip-flop

The logic symbol for the S–R flip-flop is shown in Fig. 6.1, together with the logic network of two cross-coupled NOR gates—which may be used to simulate the operation of the S–R flip-flop.

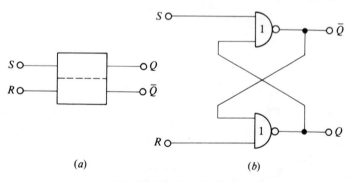

(a) *(b)*

Fig. 6.1. The S–R flip-flop.

By the definition of this flip-flop, a logical 1 signal applied to the S input causes the output Q to be *SET* to logical 1 irrespective of its previous logical state. At the same time the $\bar{Q}$ output becomes logical 0. A logical 1 signal applied to the R input causes the output Q to be *RESET* to logical 0 (and $\bar{Q}$ becomes logical 1).

Logical 1 signals simultaneously applied to S *and* R (by definition) will result in an indeterminate output state–a condition which must be

70

avoided, since the internal oscillations set up will destroy the flip-flop—if it is made using discrete components.

The truth table for this flip-flop is shown in Fig. 6.2, where Q_{t-1} represents the state of the output Q *before* the application of input signals to S and R, and Q_t represents the state of the output Q *after* the application of the given input signals to S and R.

S	R	Q_{t-1}	Q_t	
0	0	0	0	
0	0	1	1	
1	0	0	1	
1	0	1	1	
0	1	0	0	
0	1	1	0	
1	1	0	*	Indeterminate state
1	1	1	*	

Fig. 6.2. Truth table for the S–R flip-flop.

Note: The logic network using two cross-coupled 2i/p NOR gates will *NOT* be destroyed by the simultaneous application of logical 1 signals to S and R. Under these conditions *BOTH* outputs will be at logical 0 for this arrangement.

Practical exercise 6A

S–R FLIP-FLOP

Connect up the logic network of the NOR memory shown in Fig. 6.3.

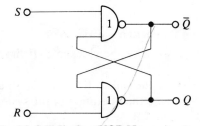

Fig. 6.3. S–R flip-flop (NOR Memory) network.

Apply logical signals to the S and R inputs to determine the truth table for this arrangement.

Repeat this exercise using two cross-coupled 2i/p NAND gates.

Note: The signals applied to the S and R inputs of the NAND network will be the complement of those required for the NOR network.

6.3 The clocked S–R flip-flop

The logic symbol and the logic network for the Clocked S–R flip-flop are shown in Fig. 6.4, and is a modification of the simple S–R flip-flop by the addition of a third input called the Clock Input C (or CK, or CLK).

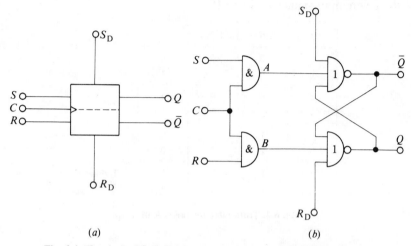

(a)   (b)

Fig. 6.4. The clocked S–R flip-flop. (a) logic symbol, (b) logic network.

When the clock input C becomes logical 1, the points A and B correspond to the points S and R respectively, so that the state of the flip-flop can only change when the clock input C is at logical 1.

Additional inputs S_D and R_D (*direct set* and *direct reset*) override the clock input, allowing the circuit to be set or reset independent of the clock input. These inputs may sometimes be referred to as *PRESET* and *PRE-CLEAR*.

Practical exercise 6B

CLOCKED S–R FLIP-FLOP

Connect up the logic network of the clocked S–R flip-flop shown in Fig. 6.5.

Adjust the frequency of the pulses from the 555 pulse generator to a suitable speed and investigate the operation of the system as combinations of signals are applied to S and R.

Disconnect the pulse generator and investigate the effect on the output of applying signals to the terminals S_D and R_D.

6.4   The D-type flip-flop (latch)

This type of flip-flop was developed to overcome the limitation of the S–R flip-flop of indeterminate operation when both S and R are at logical 1

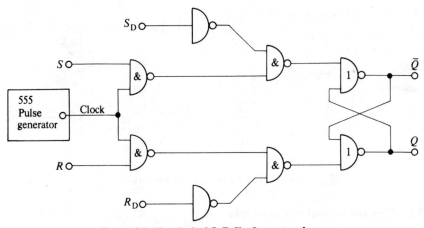

Figure 6.5. The clocked S–R flip-flop network.

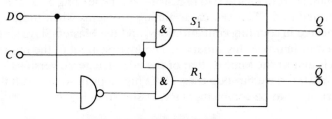

Fig. 6.6. The D-type flip-flop.

simultaneously. It achieves this by ensuring that the S and R inputs are always complementary. The simple logic network of the D-type flip-flop is shown in Fig. 6.6.

The truth table for the D-type flip-flop is shown in Fig. 6.7.

D	Q_{t-1}	Q_t
0	0	0
0	1	0
1	0	1
1	1	1

Fig. 6.7. Truth table for the D-type flip-flop.

Practical exercise 6C

D-TYPE FLIP-FLOP

Connect up the logic network of the D-type flip-flop shown in Fig. 6.8, and verify its operation.

73

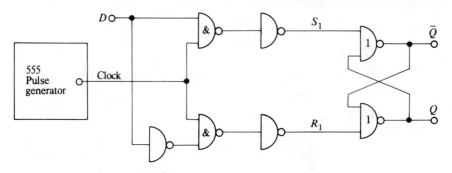

Fig. 6.8. The D-type flip-flop network.

6.5 The master–slave principle

The clocked S–R flip-flop and the D-type flip-flop are *edge-triggered* systems, i.e., the set or reset operation is initiated as soon as the clock signal changes from logical 0 to logical 1—on the leading edge of a positive pulse applied to C. This can cause timing problems when these elements are connected in counting configurations, and the Master–Slave system was developed to improve this situation. The logic symbol of the master–slave flip-flop is almost the same as that of the edge-triggered versions.

The master–slave flip-flop uses *TWO* flip-flops in series, with the clock inputs arranged to be complementary, as shown in Fig. 6.9.

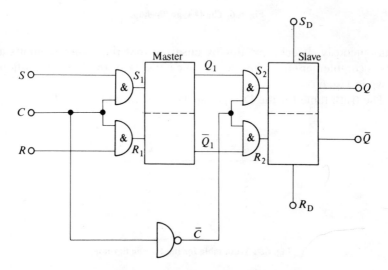

Fig. 6.9. The S–R master-slave flip-flop.

The master flip-flop can only be set or reset when the clock input C is logical 1, i.e., on the leading edge of the pulse applied to C. At the same time the slave flip-flop is inoperative, since its clock input is at logical 0.

74

When the clock input changes to logical 0, i.e., on the trailing edge of the pulse applied to C, the slave flip-flop is forced to change its state to that of the master output, since $\bar{C}$ is now at logical 1. Thus, the system is 'primed' on the leading edge of the clock pulse when the master is set (or reset). On the trailing edge of the clock pulse, the logical state of the master is transmitted to the output of the slave, so that the transition of an input signal to the master–slave flip-flop is completed when the clock pulse is on its trailing edge.

6.6 The J–K flip-flop

When flip-flops are used as counting elements some feedback is involved from the output to the input. With edge-triggered systems there is likely to be oscillation between one state and the other as long as the clock input remains at logical 1. Hence the requirement for the master–slave system.

The J–K flip-flop combines the capabilities of the clocked S–R flip-flop with the master–slave principle as shown in Fig. 6.10.

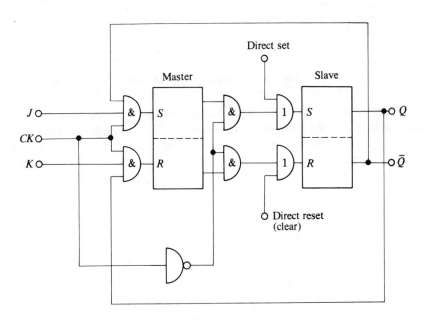

Fig. 6.10. The J–K flip-flop.

The truth table for the J–K flip-flop is shown in Fig. 6.11.

When the signal applied to J = K = logical 1 this circuit operates as a TRIGGER flip-flop, whose output Q changes state at the *end* of each clock pulse as shown in Fig. 6.12.

75

J	K	Q_{t-1}	Q_t
0	0	0	0
0	0	1	1
0	1	0	0
0	1	1	0
1	0	0	1
1	0	1	1
1	1	0	1
1	1	1	0

Fig. 6.11. Truth table for the J–K flip-flop.

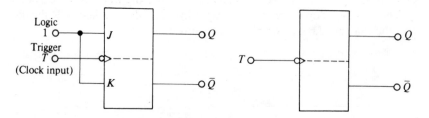

Fig. 6.12. The trigger T-type flip-flop.

The J–K flip-flop is the most versatile, since most of the desired facilities can be provided from clocked bistables. The D-type latch is the most common edge-triggered device and is readily available in various IC configurations.

7 Binary arithmetic processes

7.1 Introduction

The four most important arithmetic operations—addition, subtraction, multiplication and division—can all be performed by manipulations of addition and for this reason addition is a very important operation. The circuits used to perform the function of addition fall into two groups—half-adders and full-adders.

A half-adder produces the sum of two binary digits together with a carry digit if necessary. Each stage of a full-adder can handle one digit of each of two numbers to be added *and* the carry digit from the previous stage. Adders can be designed to operate in either a *serial* or *parallel* mode. In the serial mode, addition occurs sequentially, starting with the least significant digit (LSD). In the parallel mode, addition of all the digits is effected simultaneously. Parallel adders perform the total addition operation much more quickly than serial adders but are more complex and therefore more expensive.

7.2 Binary addition

The same basic mathematical processes are involved in the addition of two numbers regardless of the radix of the numbering system.

Starting with the least significant digit (LSD), i.e., on the extreme right, add the 1's in the usual way:

(a) If the number of 1's is EVEN, write 0 in the total; divide the ACTUAL number of 1's by 2 and carry the quotient to the next significant digit column.

(b) if the number of 1's is ODD, write 1 in the total; subtract 1 from the ACTUAL number of 1's, divide the remainder by 2 and carry the quotient to the next significant digit column.

Example 7.1

Consider the binary addition of the two denary numbers 46 and 22.

$$
\begin{array}{rcl}
101110 &\equiv& 46 \\
10110 &\equiv& 22 \\
\hline
1000100 &\equiv& 68
\end{array}
$$

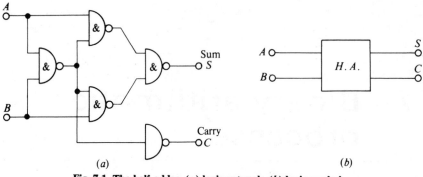

Fig. 7.1. The half-adder. (*a*) logic network, (*b*) logic symbol.

Logic networks dealing with binary addition are based on the *half-adder* as shown in Fig. 7.1. This is essentially an exclusive-OR network with an additional gate, and is capable of dealing with the sum of two binary digits.

The truth table for the half-adder is shown in Fig. 7.2.

A	B	S	C
0	0	0	0
0	1	1	0
1	0	1	0
1	1	0	1

Fig. 7.2. Truth table for the half-adder.

The Boolean equations describing the operation of the half-adder are therefore given by:

$$S = A.\bar{B} + \bar{A}.B$$

and

$$C = A.B$$

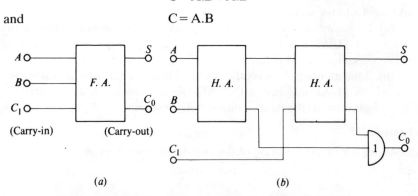

Fig. 7.3. The full-adder. (*a*) logic symbol, (*b*) using two half-adders.

The half-adder is only capable of adding the LSD's of *two* numbers, but cannot cope with the other columns of the numbers to be added since a carry digit may have been produced from the previous column. The *full-adder* was developed to cater for this requirement. The full-adder has *three* inputs, and can be made up of two half-adders as shown in Fig. 7.3.

The full-adder may be represented by the logic network shown in Fig. 7.4 using 2 i/p NAND gates.

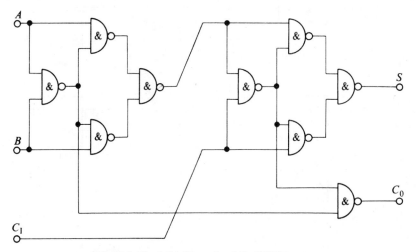

Fig. 7.4. The full-adder using 2 i/p NAND gates.

Practical exercise 7A

THE HALF-ADDER

Connect up the logic network of the half-adder as shown in Fig. 7.1, using $1 \times$ SN 7400, QUAD 2 i/p NAND gate, and $\frac{1}{6} \times$ SN 7404, Hex Inverter.

Apply all combinations of logical input signals to A and B to determine the sum S and carry C output states given in the truth table shown in Fig. 7.2.

Practical exercise 7B

THE FULL-ADDER

Connect up the logic network of the full-adder shown in Fig. 7.4, using $2\frac{1}{4} \times$ SN 7400, QUAD 2 i/p NAND gates.

Apply all combinations of logical input signals to A, B and C_I to determine the truth table for the network.

7.3   The serial adder

The full-adder considered above is capable of dealing with the addition of one column of two binary numbers, together with a carry digit from the previous column. We will now consider the complete process of adding two binary numbers.

In *serial* addition, the digits of the binary numbers to be added are presented to the adder as a timed sequence from, say, the store of a computer. One simple arrangement of a serial adder is shown in Fig. 7.5.

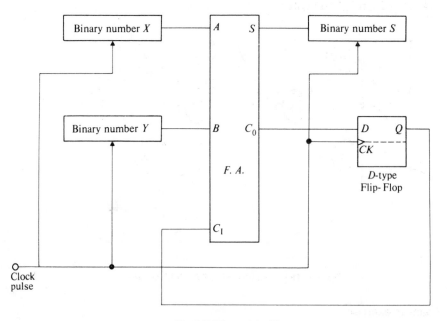

Fig. 7.5. The serial adder.

The binary numbers X, Y and the sum S are held in stores called *shift registers* (see chapter 8). Each digit of the numbers X and Y is sequentially *shifted* into the adder under the control of a clock pulse (CLK), starting with the LSD. By using a common clock pulse to control the data flow into the full-adder, sum register and operation of the D-type flip-flop, the required time delay is achieved for the transfer of the carry digit through the flip-flop to the carry-in (C_I) terminal of the full-adder.

Assuming that the Q output of the D-type flip-flop is initially at logical 0, then on application of the first clock pulse the LSD's of the two binary numbers X and Y are presented to the A and B inputs of the full-adder. The sum S appears at the output of the full-adder and the carry digit is applied to the D-type flip-flop. On the next clock pulse, the next significant digits are presented to A and B, the previous sum S output is shifted into the S register, the carry digit is propagated through to the Q output of the flip-flop and applied to the carry-in, C_I, of the full-adder. This procedure is repeated until all the digits of both numbers X and Y have been dealt with, and the sum S is held in its register.

7.4 The parallel adder

Parallel addition is capable of producing an answer almost instantaneously. However, to achieve this, additional equipment must be used. The *parallel* adder requires a full-adder for each digit of the binary numbers to be added. One simple form of parallel adder is shown in Fig. 7.6, which is capable of adding two four-digit binary numbers, i.e., a *quad full adder.* It is important to note that it is necessary to clear the carry-out digit *before* any new calculation is performed.

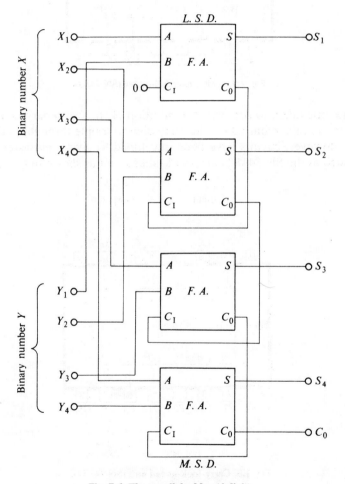

Fig. 7.6. The parallel adder (4 digit).

This type of arrangement is available in IC form, the SN 7483 being a 4 bit full-adder as shown in Fig. 7.7, labelled for the addition of two binary numbers designated A and B, and the symbol Σ (Greek 'sigma') represents the 'sum'.

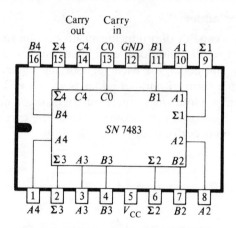

Fig. 7.7. 4-bit binary full-adder (SN 7483).

The time taken to complete the addition of the two binary numbers is equal to the time required for the carry digit to 'ripple through' to the final stage. However, circuits have been developed to give even faster operation, such as the SN 74182 carry-look-ahead unit as shown in Fig. 7.8.

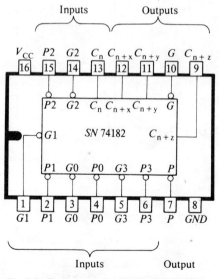

Fig. 7.8. Carry-look-ahead unit (SN 74182).

7.5 Binary subtraction

Several methods are available for the subtraction of two binary numbers. We shall consider two of these methods—the arithmetic method and the machine (computer) method:

(a) *Arithmetic method.* Starting with the LSD's of the two numbers, work a column at a time towards the MSD column:

   (i) $0-0$, and $1-1$, write 0 in the total,

  (ii) $1-0$, write 1 in the total,

 (iii) $0-1$, write 1 in the total, and in each column of the larger number (working towards the MSD) change 0 to 1 until a 1 is reached. Change this 1 to 0, and continue with the subtraction.

Example 7.2

Consider the binary subtraction of the two denary numbers, 40 minus 22:

$$
\begin{array}{r}
0\ \ \ 0\ 1 \\
\cancel{1}\ 0\ \cancel{1}\ \cancel{0}\ 0\ 0 \equiv 40 \\
\underline{1\ 0\ 1\ 1\ 0} \equiv 22 \\
\underline{1\ 0\ 0\ 1\ 0} \equiv 18
\end{array}
$$

(b) *Machine method.* Negative numbers are generally represented by writing a *minus* sign in front of them. However, there is generally no direct way by which a minus sign can be stored in a digital computer—it is only possible to store 0's and 1's. But, we *could* use an additional binary digit to represent the mathematical sign of the number—0 for positive (or plus), and 1 for negative (or minus). However, it is essential to state the number of digits in the number, to be able to use this technique, and the 'sign' digit is placed in front of the number as the MSD, e.g., *'plus 22'* is represented by 010110 (using a *total* of SIX bits in the number, and *'minus 22'* is represented by 110110. This is referred to as the *signed modulus* system, which enables us to *represent* the number, but will not enable the machine to perform subtraction.

Subtraction of two binary numbers is most generally achieved by forming the *two's complement* of the smaller number, and then adding.

The *two's complement* of the number to be subtracted is formed by firstly ensuring that this number has the same number of digits as the larger number. Secondly, all the 1's are changed to 0's, and all the 0's are changed to 1's. Finally, a 1 is added to the LSD

Example 7.3

Consider the binary subtraction of the two denary numbers 40 and 22:

$$40_{10} - 22_{10} = 101000_2 - 10110_2$$

Firstly: $10110 = 010110$

Secondly: Two's complement of $010110 = 101001$

Thirdly: Plus 1 $101001 +$

$$
\begin{array}{r}
\underline{1} \\
101010
\end{array}
$$

Finally: add: 101000

$$
\begin{array}{r}
\underline{+101010} \\
1010010
\end{array}
$$

But, since there are only SIX digits in this system, the logical 1 in the extreme left hand column is an overflow digit, i.e., it is lost—and the solution is $010010 = 10010_2 \equiv 18_{10}$, as before.

7.6 The binary subtractor

We can subtract a binary number Y from a binary number X by adding the two's complement of Y to the binary number X. The two's complement can be generated by presenting each bit of the number Y to an inverter (NOT gate), and to add a 1 to the LSD. It is therefore quite a straightforward procedure to modify the adder networks previously considered to enable them to perform subtraction.

The logic network arrangement shown in Fig. 7.9 enables two 4-bit binary numbers to be added *or* subtracted in parallel depending on the logical signal applied to the control line—logical 0 for addition, and logical 1 for subtraction.

Practical exercise 7C

THE 4-BIT PARALLEL ADDER

Connect up the logic arrangement of the 4-Bit Parallel Adder as shown in Fig. 7.6 using $1 \times$ SN 7483 quad full-adder shown in Fig. 7.9.

Set the 4-Bit binary numbers X and Y to the required values. Switch on the supply, and monitor the Sum outputs together with the carry-out digit. Check the answer by arithmetic methods.

Repeat with different values of X and Y.

Practical exercise 7D

THE 4-BIT PARALLEL ADDER/SUBTRACTOR

Connect up the logic arrangement of the 4-Bit parallel adder/subtractor shown in Fig. 7.9, using $1 \times$ SN 7483, quad full-adder, $3 \times$ SN 7400, Quad 2 i/p NAND, $\frac{5}{6} \times$ SN 7404, Hex inverter.

Set up the 4-Bit binary numbers X and Y to the required values. Switch on the supply, and with the Control signal set to logical 0, monitor the output states and confirm that the output is the *sum* of X and Y. Now, set the Control signal to logical 1 and confirm that the output is the difference between X and Y, i.e., $X - Y$.

Repeat with different values of X and Y.

7.7 Binary multiplication

The standard methods of multiplication may be applied, but since we only ever need to multiply by 0 and by 1, multiplication becomes a process of adding and shifting.

84

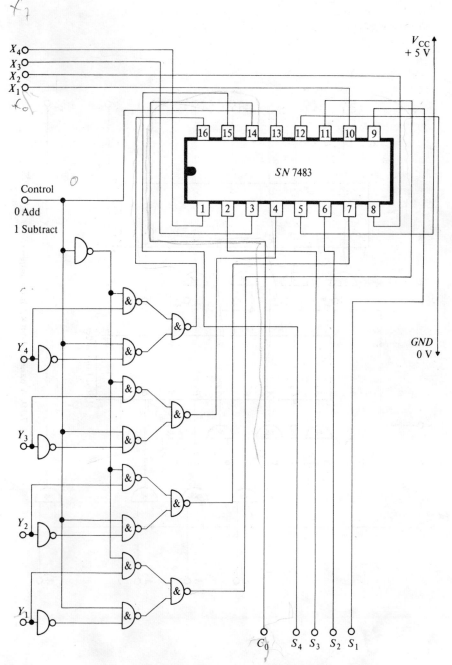

Fig. 7.9. 4-bit full-adder/subtractor.

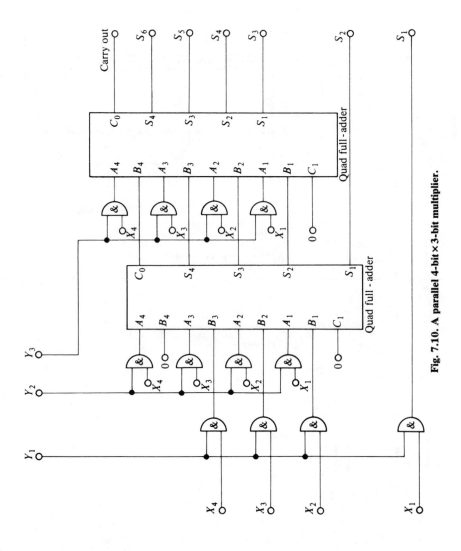

Fig. 7.10. A parallel 4-bit × 3-bit multiplier.

Example 7.4

Consider the binary multiplication of the two denary numbers 11 and 5.

$$
\begin{array}{r}
1011 \equiv 11 \\
\times 101 \equiv 5 \\
\hline
1011 \\
1011 \\
\hline
110111 \equiv 55
\end{array}
$$

A logic network capable of multiplying a 4-Bit binary number X by a 3-Bit binary number Y is shown in Fig. 7.10.

Practical exercise 7E

THE PARALLEL MULTIPLIER

Connect up the logic arrangement of the parallel multiplier shown in Fig. 7.10, using $2 \times$ SN 7483, quad full-adder, $3 \times$ SN 7400, Quad 2 i/p NAND gates, $2 \times$ SN 7404, Hex inverter.

Set up a 4-Bit binary number to the X inputs and a 3-Bit binary number to the Y inputs. Switch on the supply and monitor the output states. Evaluate the product of X and Y by arithmetic methods and compare the results.

Repeat with different values of X and Y.

7.8   Binary division

Here again, the standard mathematical techniques may be applied.

Example 7.5

Consider the binary division of the two denary numbers 55 and 5.

$$
\begin{array}{ll}
\underline{1011}\ \text{Quotient} \qquad\quad \underline{11}\ \text{Quotient} \\
101\overline{)110111} \qquad\qquad\quad 5\overline{)55} \\
\underline{101} \qquad\qquad\qquad\quad\ \underline{5} \\
111 \qquad\qquad\qquad\quad\ 5 \\
\underline{101} \qquad\qquad\qquad\quad\ \underline{5} \\
101 \qquad\qquad\qquad\ \ \ 0\ \ \text{Remainder} \\
\underline{101} \\
000\ \ \text{Remainder}
\end{array}
$$

Division is therefore achieved by a process of subtracting and shifting, and logic networks to achieve this can be devised using similar techniques to those previously described.

7.9   The rate multiplier

The rate multiplier is a very useful arrangement whose output is made up of a train of pulses whose average rate of repetition depends on the product

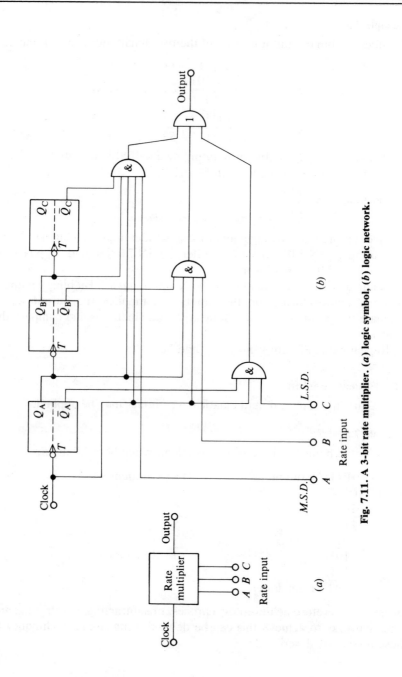

Fig. 7.11. A 3-bit rate multiplier. (a) logic symbol, (b) logic network.

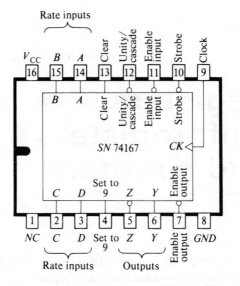

Fig. 7.12. 4-bit decade rate multiplier (SN 74167).

of the repetition rate of the input pulses and the parallel binary input as shown in Fig. 7.11.

This type of arrangement is available in IC form, the SN 74167 being a 4-Bit decade rate multiplier as shown in Fig. 7.12.

With some slight modifications, the rate multiplier can be arranged to perform several mathematical functions, including: addition, subtraction, multiplication, division, exponentials, integration and other more complex functions. Other arrangements which include the mathematical functions detailed above may be included in IC packages and listed as arithmetic logic units (ALU's).

8 Registers, shift registers and counters

8.1 Introduction

When a group of flip-flops are connected together so that they store related information, they are known collectively as a *register*.

A *shift register* is one which is designed so that data may be shifted along the register either to the right or to the left.

Certain types of register can be used for the purpose of counting pulses, and are known as *counters*.

A *ring counter* is a shift register which is connected in the form of a continuous ring.

8.2 The storage register

Storage registers can be made up by using virtually any of the flip-flops previously described in chapter 7, but those which are controlled by clock pulses are most common. A storage register is required to accept data in from a suitable source, store that data and hold it ready for use at a later time.

A simple 3-bit storage register using D-type flip-flops is shown in Fig. 8.1, which is a *parallel-in* and *parallel-out* (PIPO) register.

In this simple register, the data to be stored is presented as a binary number to the D inputs of the flip-flops (all bits simultaneously), and on application of a clock pulse, the data is stored in the register and is readily available at the Q outputs as shown. Flip-flops having *preset* and *preclear* facilities such as the SN 7474 Dual D-type flip-flop, may use these connections to feed data in to the register.

Note: With the SN 7474, if no input is applied to the D inputs (i.e., it is open-circuit), then it normally assumes a logical 1 signal level.

Also the 'preset' and 'preclear' inputs are operated by a logical 0 signal level—and these inputs override both the D inputs and the clock input.

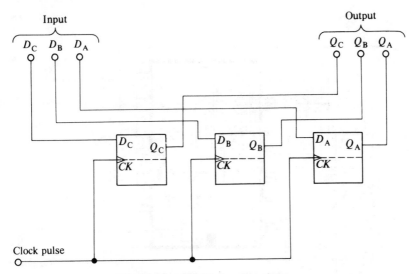

Fig. 8.1. Simple 3-bit storage register.

Practical exercise 8A

4-BIT STORAGE REGISTER

Connect up the storage register shown in Fig. 8.2 and Fig. 8.3, using $2 \times$ SN 7474 Dual D-type flip-flops.

Apply the 4-bit binary number to the D inputs as shown (i.e., a parallel number). Set the CLOCK input to logical 1 to store the number in the register. Remove the CLOCK connection and observe the register contents indicated by the LED's (connected to the Q outputs).

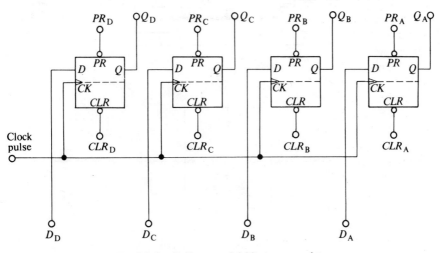

Fig. 8.2. Logic diagram of 4-bit storage register.

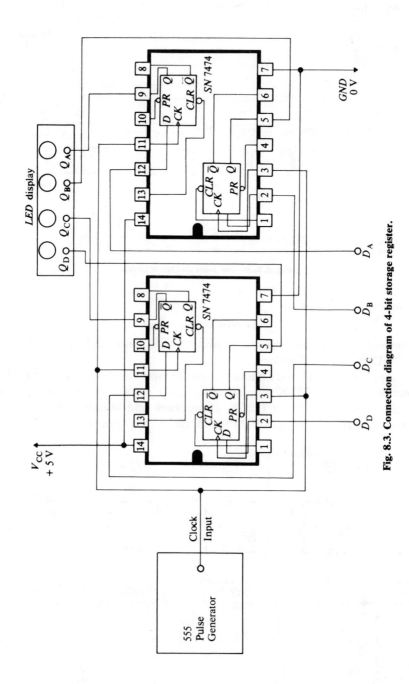

Fig. 8.3. Connection diagram of 4-bit storage register.

Reconnect the CLOCK input and set the signal to logical 1. Set the CLOCK to logical 0 and observe that the register contents are cancelled as the CLOCK signal level changes from logical 1 to logical 0.

8.3 The shift register

The simple storage register can be modified to a shift register by feeding the output of one flip-flop into the next, and so on. A simple shift-right register using D-type flip-flops is shown in Fig. 8.4 (*a*).

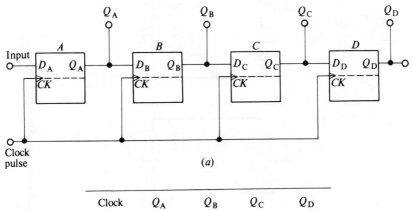

(*a*)

Clock	Q_A	Q_B	Q_C	Q_D
0	0	0	0	0
1	1	0	0	0
2	0	1	0	0
3	0	0	1	0
4	0	0	0	1
5	0	0	0	0

(*b*)

Fig. 8.4. Simple shift-right register. (*a*) Logic diagram, (*b*) truth table.

Thus if we require a logical 1 to be shifted along this register, one flip-flop at a time (i.e., at each clock pulse) then it is necessary to initially apply a logical 1 to the input D_A at the same time as the first clock pulse. At the end of the first clock pulse, Q_A becomes logical 1, and D_A must now be held at logical 0 for the remainder of the operational sequence. Each clock pulse causes the logical 1 signal to be propagated *one flip-flop to the right* as shown in the truth table in Fig. 8.4 (*b*).

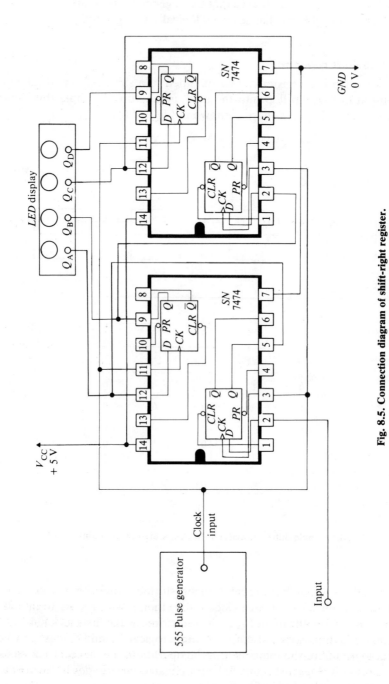

Fig. 8.5. Connection diagram of shift-right register.

Practical exercise 8B

SHIFT-RIGHT REGISTER

Connect up the shift-right register shown in Fig. 8.4 (a) and Fig. 8.5, using $2 \times$ SN 7474 dual D-type flip-flops.

Apply the CLOCK input from the 555 pulse generator, and adjust the frequency of the pulse to its slowest rate. Set the INPUT to logical 1 and observe that the output Q_A goes to logical 1 at the next clock pulse. Set the INPUT to logical 0 at this instant. Now observe that the logical 1 signal *SHIFTS TO THE RIGHT* one flip-flop at a time as the CLOCK pulses are fed in. After *four* clock pulses the register is empty.

8.4 Data control in shift registers

We have seen from the simple register and shift register that data can be fed either in parallel (all digits applied simultaneously) or in series (digits are sequentially applied to the first flip-flop). Similarly, data can be taken out from the register in parallel (all digits are read simultaneously) or in series (digits are read out sequentially from the last flip-flop).

If the flip-flops used in the register do not have preset and preclear facilities, then parallel data input can be controlled by the use of additional logic circuitry, such as the arrangement shown in Fig. 8.6.

When the *ENABLE DATA INPUT* is at logical 1, the inputs to the D-type flip-flops are the same as the required data to be stored, so that on the application of a clock pulse, the data is stored in the register. If the ENABLE DATA INPUT signal is at logical 0, the input to each flip-flop is the same as the Q output of the previous one.

A shift register having parallel-input using flip-flops with preset and preclear facilities is shown in Fig. 8.7.

It can be seen that in both registers shown in Fig. 8.6 and Fig. 8.7, the data can be taken out in parallel by 'reading' the Q output states of the flip-flops simultaneously. Furthermore, once the data has been fed in, it will be shifted one place to the *right* at each clock pulse.

The system shown in Fig. 8.7 can also be used as a serial-in/serial-out (SISO) shift register if the data is fed in sequentially to D_A and read out of Q_D.

Practical exercise 8C

PARALLEL-IN/PARALLEL-OUT 4-BIT SHIFT REGISTER

Connect up the shift register together with the control logic shown in Fig. 8.7 and Fig. 8.8, using $2 \times$ SN 7474 Dual D-type flip-flops, $2 \times$ SN 7400 Quad 2 i/p NAND gates, $\frac{2}{3} \times$ SN 7404 Hex inverter.

Apply the PARALLEL BINARY NUMBER to the input terminals A, B, C and D simultaneously. Set ENABLE PARALLEL DATA to logical 1 and observe the parallel number on the LED's connected to Q_A, Q_B, Q_C

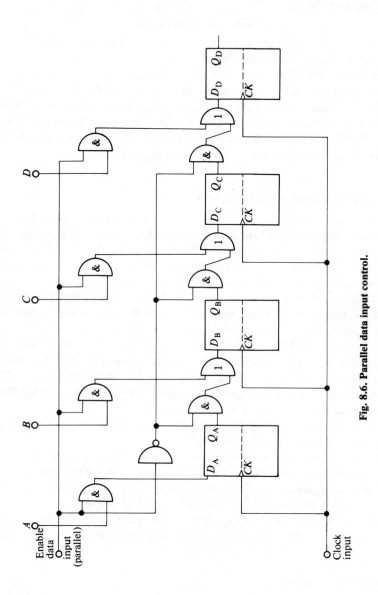

Fig. 8.6. Parallel data input control.

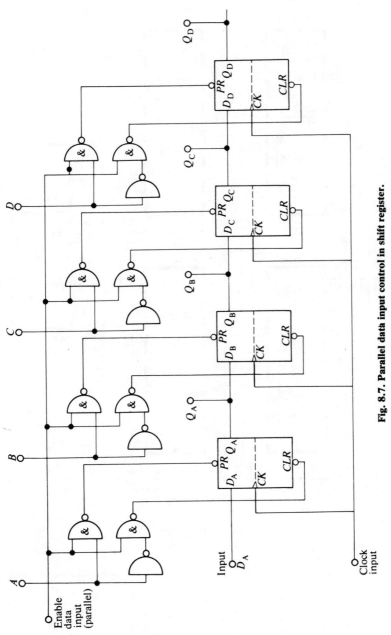

Fig. 8.7. Parallel data input control in shift register.

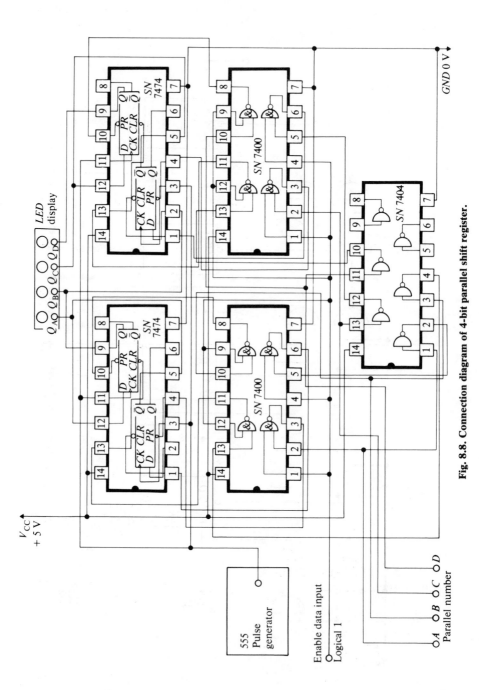

Fig. 8.8. Connection diagram of 4-bit parallel shift register.

and Q_D. Set ENABLE PARALLEL DATA to logical 0 and check that the parallel data is shifted out of the register at the same time as the CLOCK pulse is applied.

8.5   Reversible shift register

The registers considered above are only capable of shifting data to the right. However, some applications exist in which it is required that data is shifted to the right *or* to the left. This requirement can be satisfied by the use of a logic control network as shown in Fig. 8.9.

The logic diagram of a 4-bit reversible shift register having facilities for parallel input/output (PIPO) and serial input/output (SISO) is shown in Fig. 8.10 which can be constructed using $2 \times$ SN 7474 Dual D-type Flip-Flops, $2 \times$ SN 7400 Quad 2 i/p NAND gates, $2 \times$ SN7450 Dual 2 wide 2 i/p AND—OR—INVERT gates, $1\frac{1}{6} \times$ SN 7404 Hex inverter.

In the register shown in Fig. 8.10, the 4-bit binary number is applied to the *PARALLEL DATA IN* terminals. The *DATA LOAD* is set to logical 1, which stores the number in the register. Logical signals may now be applied to the *SHIFT CONTROL*, and the contents of the register will then be shifted (to the right or to the left) on application of the *CLOCK* pulses.

Alternatively, the 4-bit number may be fed in sequentially to either the *SERIAL INPUT—SHIFT LEFT* or to the *SERIAL INPUT—SHIFT RIGHT*. Again, the data will be shifted through the register on application of the *CLOCK* pulses.

Practical exercise 8D

4-BIT PARALLEL/SERIAL CONVERTER

Connect up the converter arrangement shown in Fig. 8.11, using $1 \times$ SN 7495 4-bit shift register (PIPO), $\frac{1}{2} \times$ SN 7420 Dual 4 i/p NAND gate, $\frac{1}{4} \times$ SN 7400 Quad 2 i/p NAND gate.

Apply the parallel 4-bit binary number to the *PARALLEL INPUT*, and a logical 0 signal to the *INITIAL INPUT* to set the register to the known state.

Apply *CLOCK* pulses to the *ENTER SHIFT INPUT* and monitor the *serial* output and the *parallel data*. After 4 clock pulses all the data has been shifted out.

Practical exercise 8E

4-BIT SERIAL/PARALLEL CONVERTER

Connect up the converter arrangement shown in Fig. 8.12, using $1 \times$ SN 7495 4-bit shift register (PIPO), $\frac{1}{4} \times$ SN 7402 Quad 2 i/p NOR gate, $\frac{1}{6} \times$ SN 7404 Hex inverter.

99

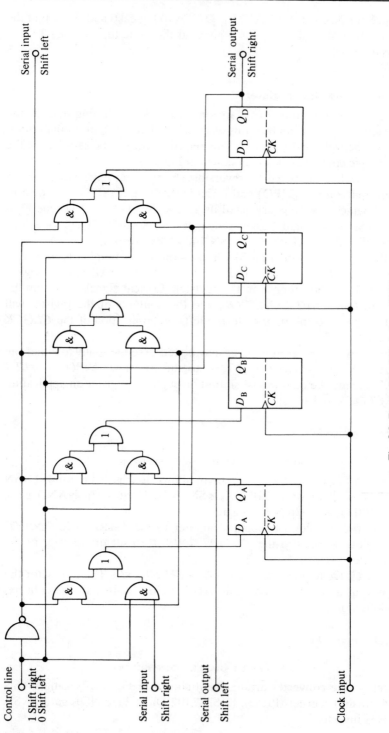

Fig. 8.9. Reversible shift register, SISO.

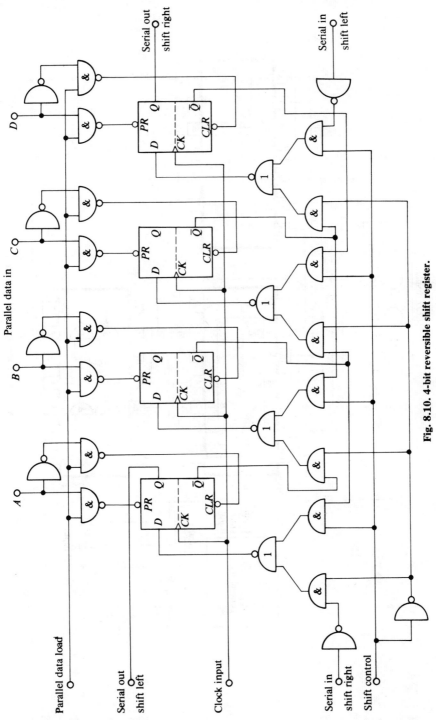

Fig. 8.10. 4-bit reversible shift register.

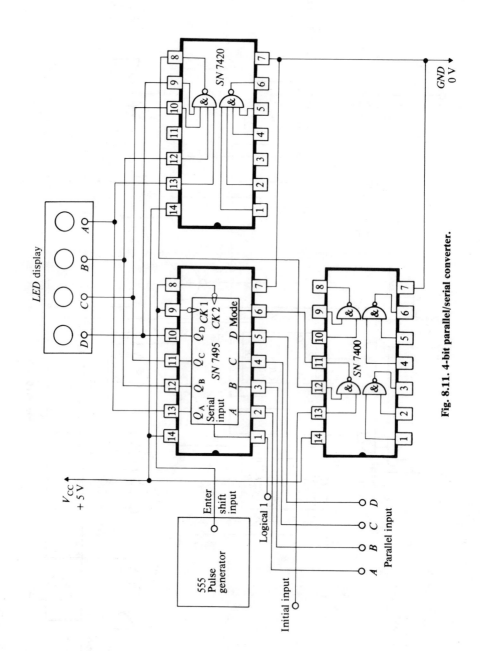

Fig. 8.11. 4-bit parallel/serial converter.

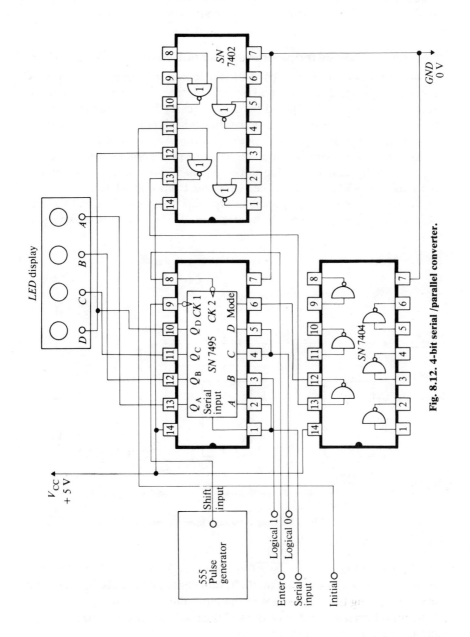

Fig. 8.12. 4-bit serial /parallel converter.

103

Apply a logical 1 signal to *ENTER*, and the 4-bit binary number in sequence to the *SERIAL INPUT* simultaneously with the *CLOCK* pulses applied to the *SHIFT* input. Monitor the parallel data output.

8.6 Asynchronous (ripple-through) counters

An asynchronous, or *serial*, counter is a sequential logic system in which the pulses are applied at one end of the counter, and the process of adding each pulse must be completed before the 'carry bit' is propagated to the following stage. This next stage must then add the carry bit to the number in that stage, i.e., the carry bit appears to '*ripple through*' the length of the counter until the count is complete. The most common method of achieving this is by using J–K flip-flops in which the J and K inputs are permanently held at logical 1, i.e., as in the T flip-flop. However, this type of counter may be set up by using D-type flip-flops as in Practical Exercise 8F. The simple ripple-through counter using J–K flip-flops is shown in Fig. 8.13.

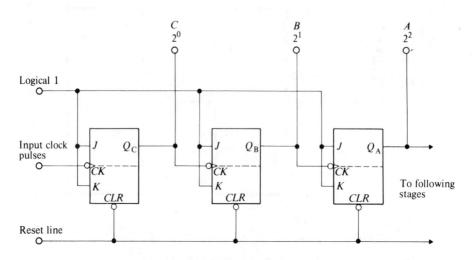

Fig. 8.13. Ripple-through binary counter.

Assuming that master-slave flip-flops are used, then each of the flip-flops shown will change state on the trailing edge of the pulse applied to its input, i.e., during the transition of the input pulse from logical 1 to logical 0. The waveforms for each of the flip-flops in response to a train of input pulses are as shown in Fig. 8.14.

This counter will count from 000_2 to 111_2 (i.e., 0_{10} to 7_{10}) and then repeat the sequence since it contains only *three* flip-flops. The count can be increased by using additional flip-flops.

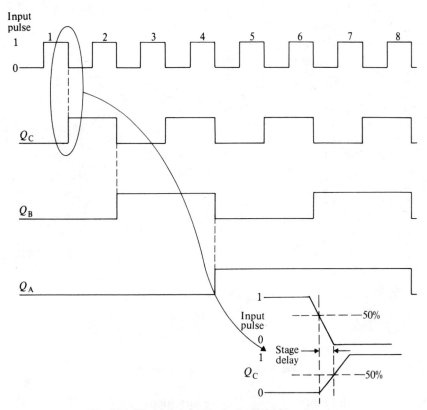

Fig. 8.14. Waveforms for ripple-through binary counter.

8.7 Reversible counter

The basic serial counter can be converted into a bi-directional counter by using logic control circuitry as shown in Fig. 8.15.

Practical exercise 8F

RIPPLE-THROUGH (SERIAL) COUNTER

Connect up the logic network of the ripple-through binary counter shown in Fig. 8.16, using $2 \times 2N$ 7474 Dual D-type flip-flops.

Adjust the frequency of the 555 pulse generator to a suitable speed and observe the binary display on the LED's.

Connect the LED's to the $\bar{Q}$ outputs of the flip-flops and observe that the display is now that of a down counter.

8.8 Decoding

Several applications exist for counters which repeat sequences which do not coincide with a power of 2, i.e., 4, 8, 16 and so on, which can be

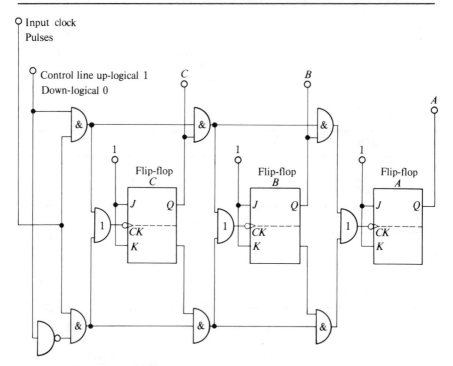

Fig. 8.15. Reversible serial counter.

arranged as described above. This requirement can be fulfilled by the relatively simple technique of using a *logic decoder*. The decoder can be made using a simple AND gate to detect the final state of the count required, e.g., the 'batch', and the output of the AND gate is applied to the RESET line, so that the count proceeds in the normal way up to the required limit then all the flip-flops are reset to zero, and the count is repeated.

However, many TTL IC flip-flops require a logical 0 signal to RESET (or CLEAR), and the decoder can therefore use a NAND gate. In such systems, as long as the RESET input is at logical 1 the flip-flop behaves as normal. A simple decoding arrangement is shown in Fig. 8.17, in which the 2 i/p NAND gate detects the state when B and C are at logical 1 when flip-flops B and C will be RESET to zero. The state detected is the state of the flip-flops *following* the final count to be displayed, i.e., in the case shown when the final count will be 101_2 (5_{10}).

Practical exercise 8G

BATCH COUNTER

Connect up the arrangement of the ripple-through counter (as used in Practical Exercise 8F) with the decoding network as shown in Fig. 8.18,

106

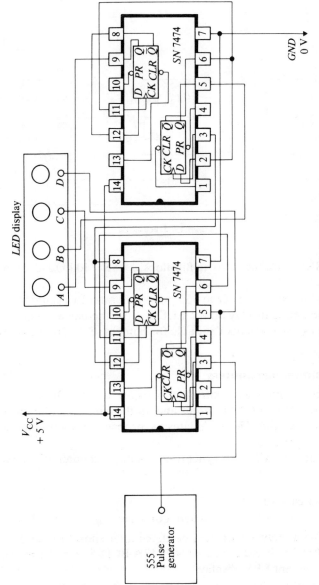

Fig. 8.16. 4-bit ripple-through binary counter.

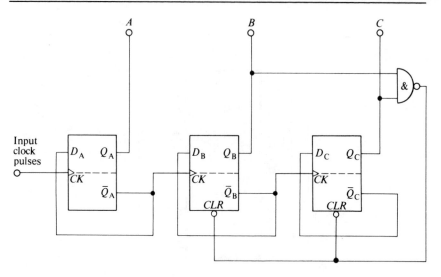

Fig. 8.17. Decoding.

using $2 \times$ SN 7474 Dual D-type flip-flops, $\frac{1}{4} \times$ SN 7400 Quad 2 i/p NAND gates.

Adjust the frequency of the 555 pulse generator to a suitable speed and observe the binary display on the LED's. Note the batch count.

Devise the logic decoder suitable for batches of different quantities.

8.9 Synchronous counters

The counting sequence is controlled by means of a CLOCK pulse applied simultaneously to all the flip-flops, and all the *changes* of all the flip-flops occur in synchronism. This eliminates the propagation delay experienced in ripple-through counters.

The logic network of a synchronous binary counter is shown in Fig. 8.19.

Practical exercise 8H

DECADE COUNTER

Connect up the network of the decade counter shown in Fig. 8.20, using $1 \times$ SN 7490 Decade counter, $1 \times$ SN 7447A BCD-Seven-segment decoder, $1 \times$ Seven-segment LED display.

Adjust the frequency of the pulses from the 555 pulse generator to a suitable speed and observe the display.

Note the effect of switching the signal applied to the R_0 (RESET to 0) terminals to logical 1 and back to logical 0.

Convert this into a batch counter (batches of 5) by connecting Q_B and Q_C to a NAND gate. The output of the NAND gate is passed through an

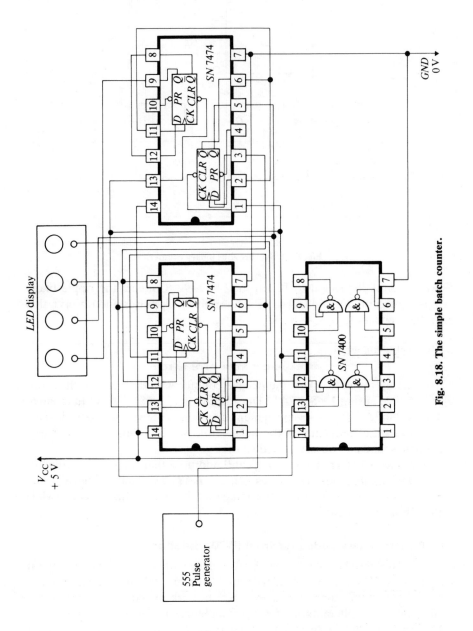

Fig. 8.18. The simple batch counter.

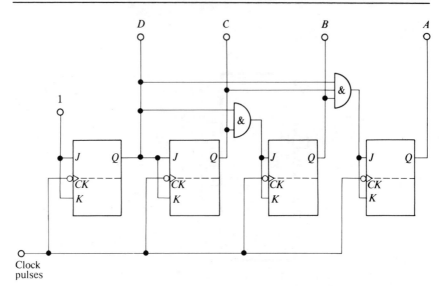

Fig. 8.19. Synchronous binary counter.

inverter and connected to R_0 on the SN 7490. Connect the ripple blank input (pin 5 on SN 7447A) to GND. Apply clock pulses and observe that display repeats 1, 2, 3, 4, 5.

Practical exercise 8l

DECADE COUNTER WITH LATCH

Connect up the network of the decade counter with the quad latch shown in Fig. 8.21, using $1 \times$ SN 7490 Decade counter, $1 \times$ SN 7475 Quad latch, $1 \times$ SN 7447A BCD—Seven segment decoder, $1 \times$ Seven segment display.

Adjust the frequency of the pulses from the 555 pulse generator to a suitable speed and observe the display.

Set the *LATCH* signal to logical 0 and note that the display *holds* its value, whilst the count continues. Set the *LATCH* signal to logical 1 and note that the display *jumps* to the value that the counter has reached (while the display was 'held').

8.10 The binary coded decimal (BCD) notation

The binary number system is the simplest and best system for digital computers. However, the denary number system is the most familiar world wide. Hence, for computers to, be able to work in the binary system we must have a simple method of converting binary to denary and vice-versa. The conventional method using powers of 2 is awkward, and although computers can be instructed to perform the conversions, human operators find it time consuming to convert long strings of binary digits into a denary number.

110

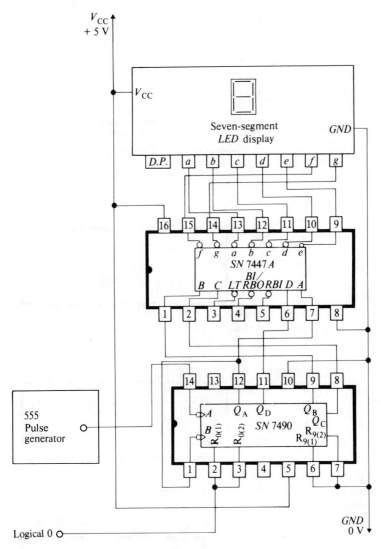

Fig. 8.20. The decade counter.

The octal system is a shorthand way of writing binaries—but it is not a great deal of help in converting them to denary.

To overcome these problems, several binary codes have been devised to *translate* EACH denary digit separately into an *equivalent 4-bit binary combination* and vice-versa, some of which are shown in the table in Fig. 8.22.

After a count of 9_{10} the BCD values (or *weights*) change by a factor equivalent to denary 10.

111

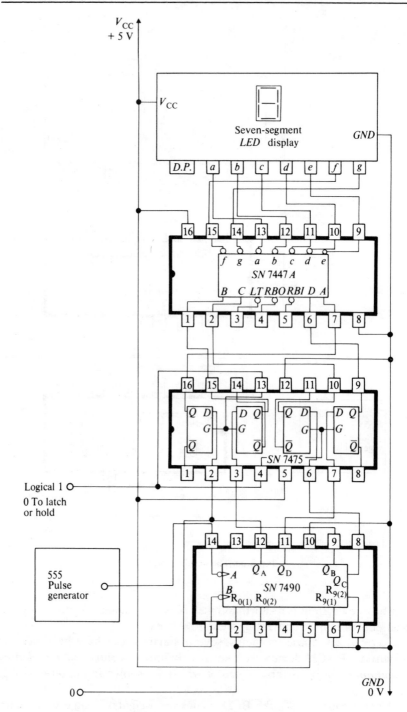

Fig. 8.21. The decade counter with latch.

Denary	Excess 3 (XS-3)	8421 BCD	2421 BCD	7421 BCD
0	0 0 1 1	0 0 0 0	0 0 0 0	0 0 0 0
1	0 1 0 0	0 0 0 1	0 0 0 1	0 0 0 1
2	0 1 0 1	0 0 1 0	0 0 1 0	0 0 1 0
3	0 1 1 0	0 0 1 1	0 0 1 1	0 0 1 1
4	0 1 1 1	0 1 0 0	0 1 0 0	0 1 0 0
5	1 0 0 0	0 1 0 1	0 1 0 1	0 1 0 1
6	1 0 0 1	0 1 1 0	0 1 1 0	0 1 1 0
7	1 0 1 0	0 1 1 1	0 1 1 1	1 0 0 0
8	1 0 1 1	1 0 0 0	1 1 1 0	1 0 0 1
9	1 1 0 0	1 0 0 1	1 1 1 1	1 0 1 0

Fig. 8.22. Binary coded decimal (BCD) systems.

Example 8.1

The denary number 79_{10} may be represented in 8421 BCD by 0111 1001 thus:

80	40	20	10	8	4	2	1
0	1	1	1	1	0	0	1

8.11 8421 BCD counter

The logic network of a common type of ripple-through 8421 BCD up-counter is shown in Fig. 8.23, and its sequence of operation follows the code shown in the table above.

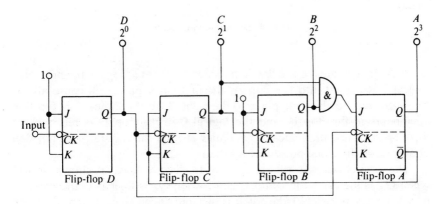

Fig. 8.23. Ripple-through 8421 BCD counter.

Flip-flops B and D are arranged as simple T flip-flops. Initially, with all the flip-flops reset to zero, the $\bar{Q}_A$ output is at logical 1 which is fed back to flip-flop C, and it too operates as a T flip-flop initially. Under these conditions, the output from the AND gate is logical 0, so that flip-flop A cannot operate. Thus, flip-flops D, C and B operate as conventional T flip-flops which count 'up' in normal binary sequence for the first seven pulses. After the seventh pulse, the output of the AND gate is logical 1, and since Q_D is also at logical 1, flip-flop A is caused to change its state by the eighth pulse and Q_B, Q_C and Q_D fall to logical 0. As soon as Q_A becomes logical 1, then $\bar{Q}_A$ is at logical 0 and flip-flops C and B cannot operate. The ninth pulse causes Q_D to become logical 1, and the tenth pulse causes Q_D to become logical 0, and simultaneously applies logical 0 to the CLOCK input of flip-flop A thus resetting Q_A to logical 0.

8.12 The ring counter

The ring counter is very basically a shift register whose input is obtained from its output, as shown in Fig. 8.24 using D-type flip-flops.

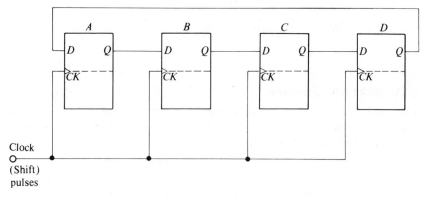

Fig. 8.24. The simple ring counter.

If all the Q states of the flip-flops are initially set at logical 0, and then a logical 1 is loaded into flip-flop A, the first clock pulse will cause all the flip-flop states to shift one place to the right. Thus the logical 1 in flip-flop A appears in flip-flop B, and the logical 0 in flip-flop D is fed back to flip-flop A. Thus, the logical 1 *circulates* around the register, one flip-flop at a time, as the clock pulses are applied.

The ring counter is relatively easy to decode, e.g., as shown above, when Q_A is logical 1 the *count* is zero, when Q_B is logical 1, the *count* is one, when Q_C is logical 1 the *count* is two, etc. It can be seen therefore that the *cycle length* of the code generated is four for the system shown above,

114

and *ten* flip-flops would be required for a decade counter. The basic ring counter is therefore not very economical in the use of flip-flops.

The cycle length of the ring counter may be *doubled* by feeding back the output from $\bar{Q}_D$ (instead of Q_D), which is then called a *twisted ring counter*. However, this is more difficult to decode as shown in the truth table in Fig. 8.25, *assuming* that the Q states of all the flip-flops are initially set at logical 0.

Denary	Q_A	Q_B	Q_C	Q_D	
0	0	0	0	0	
1	1	0	0	0	
2	1	1	0	0	
3	1	1	1	0	
4	1	1	1	1	
5	0	1	1	1	
6	0	0	1	1	
7	0	0	0	1	Repeat

Fig. 8.25. Truth table for a twisted ring counter.

To ensure that the flip-flops are initially set to the required conditions, a logical network may be used as shown in Fig. 8.26, which feeds a logical 1 signal to the input of flip-flop A when the Q states of ALL the flip-flops are at logical 0.

Practical exercise 8J

4-STAGE RING COUNTER

Connect up the arrangement of the even-cycle length ring counter shown in Fig. 8.27, using $2 \times$ SN 7474 Dual D-type flip-flops, $\frac{1}{2} \times$ SN 7420 Dual 4 i/p NAND gates, $\frac{1}{6} \times$ SN 7404 Hex inverter.

Apply a train of pulses and observe the Q states of the flip-flops on the LED's. Draw up the truth table for a complete sequence of the counter.

Practical exercise 8K

JOHNSON COUNTER (TWISTED RING)—EVEN-CYCLE LENGTH

Connect up the arrangement of the six-stage even-cycle length Johnson counter shown in Fig. 8.28, using $3 \times$ SN 74107 Dual J–K flip-flops, $\frac{1}{3} \times$ SN 7410 Triple 3 i/p NAND gates, $\frac{1}{6} \times$ SN 7404 Hex inverter.

Apply a train of clock pulses and observe the Q states of the flip-flops on the LED's. Draw up the truth table for a complete sequence of 12 pulses and compare with the truth table shown in Fig. 8.29.

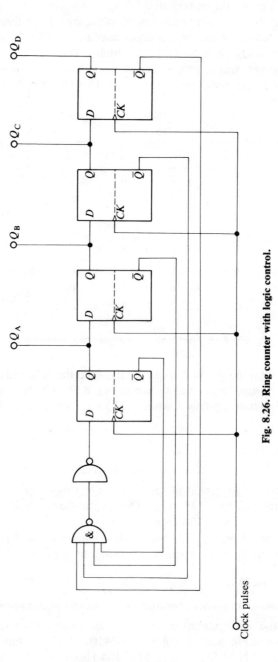

Fig. 8.26. Ring counter with logic control.

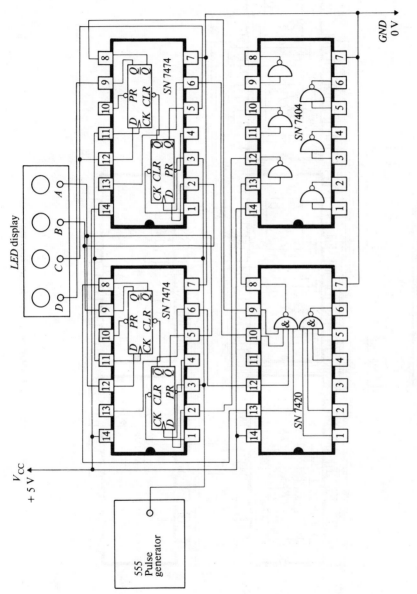

Fig. 8.27. 4-stage even-cycle length ring counter.

117

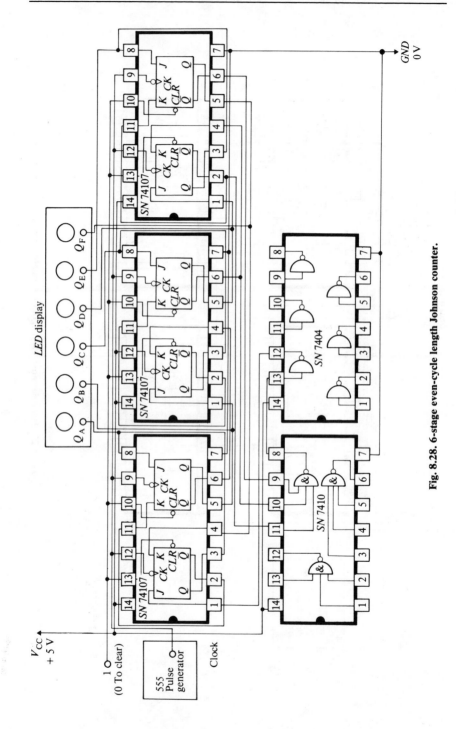

Fig. 8.28. 6-stage even-cycle length Johnson counter.

Denary	Q_A	Q_B	Q_C	Q_D	Q_E	Q_F
0	0	0	0	0	0	0
1	1	0	0	0	0	0
2	1	1	0	0	0	0
3	1	1	1	0	0	0
4	1	1	1	1	0	0
5	1	1	1	1	1	0
6	1	1	1	1	1	1
7	0	1	1	1	1	1
8	0	0	1	1	1	1
9	0	0	0	1	1	1
10	0	0	0	0	1	1
11	0	0	0	0	0	1

Fig. 8.29. Truth table for even-cycle length Johnson counter.

Practical exercise 8L

JOHNSON COUNTER (TWISTED RING)—ODD-CYCLE LENGTH

The Johnson counter can be converted into an *odd-cycle length* by using logic decoding circuitry to 'jump' the state when all the Q outputs are at logical 1. For example, in the truth table above, we would need to detect the states equivalent to a count of denary 5, and jump immediately to the states equivalent to a count of denary 7.

Connect up the arrangement of the six-stage odd-cycle length Johnson counter shown in Fig. 8.30, using $3 \times$ SN 74107 Dual J–K flip-flops, $1 \times$ SN 7430 8 i/p NAND gate, $\frac{2}{3} \times$ SN 7410 Triple 3 i/p NAND gate, $\frac{1}{6} \times$ SN 7404 Hex inverter.

Apply a train of clock pulses and observe the Q states of the flip-flops on the LED's. Draw up the truth table for a complete sequence of 11 clock pulses.

8.13 The random number generator

The random number generator is a shift register which generates a code which does not follow any logical pattern, i.e., the binary numbers are produced in a *random* sequence.

This requirement is generally satisfied by using a shift register whose input is obtained from a complex logical network which derives its signals from the flip-flops within the register.

A simplified arrangement is shown in Fig. 8.31, which does not generate a truly random sequence, and is therefore referred to as a *pseudo-*

119

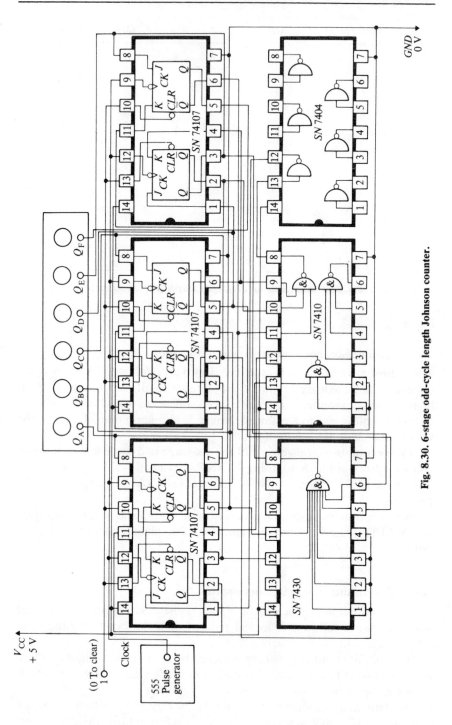

Fig. 8.30. 6-stage odd-cycle length Johnson counter.

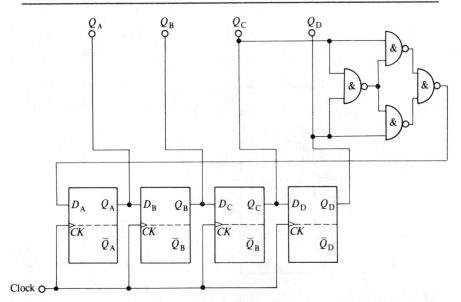

Fig. 8.31. Pseudo-random number generator.

Denary	Q_A	Q_B	Q_C	Q_D
0	1	1	1	1
1	0	1	1	1
2	0	0	1	1
3	0	0	0	1
4	1	0	0	0
5	0	1	0	0
6	0	0	1	0
7	1	0	0	1
8	1	1	0	0
9	0	1	1	0
10	1	0	1	1
11	0	1	0	1
12	1	0	1	0
13	1	1	0	1
14	1	1	1	0

Fig. 8.32. Truth table for pseudo-random number
generator.

121

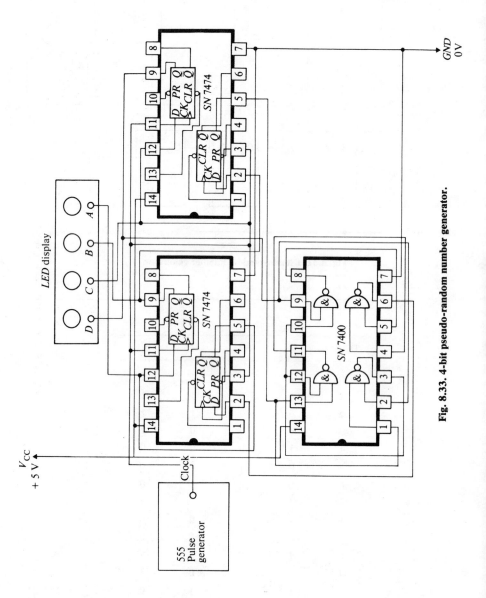

Fig. 8.33. 4-bit pseudo-random number generator.

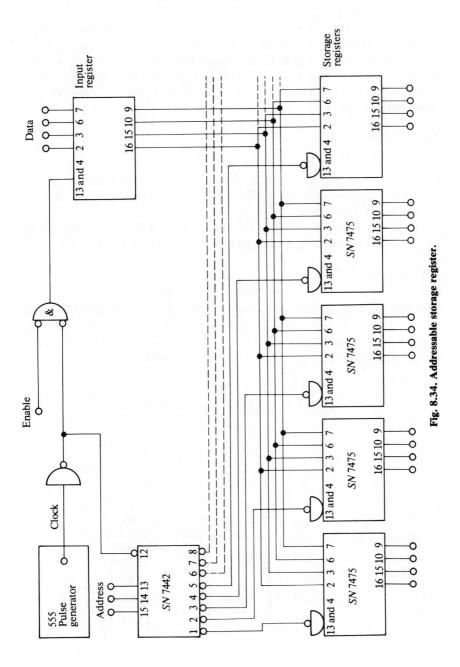

Fig. 8.34. Addressable storage register.

random number generator, i.e., if the initial state is known, the counting sequence *can* be predicted.

The logic network ensures that the counter starts with flip-flop A set at logical 1. The exclusive-OR network is used to compare the output states of flip-flops C and D and will feed a logical 1 signal to D_A when Q_C and Q_D are different, and a logical 0 signal to D_A when Q_C and Q_D are the same, so that the truth table is as shown in Fig. 8.32, which will be repeated every 15 clock pulses.

Practical exercise 8M

PSEUDO-RANDOM NUMBER GENERATOR

Connect up the arrangement of the pseudo-random number generator shown in Fig. 8.33, using $2 \times$ SN 7474 Dual D-type flip-flops, $1 \times$ SN 7400 Quad 2 i/p NAND gates.

Apply the CLOCK pulses and observe the 4-bit binary number on the LED's. Draw up the truth table and compare with that shown in Fig. 8.32.

Practical exercise 8N

ADDRESSABLE REGISTER

The SN 7475 Quad latch may be used as a 4-bit per byte addressable storage register by connecting several of them as shown in Fig. 8.34, using the SN 7442 BCD decimal decoder as the address interfacing, which can decode either a three-to-eight or a four-to-ten line register.

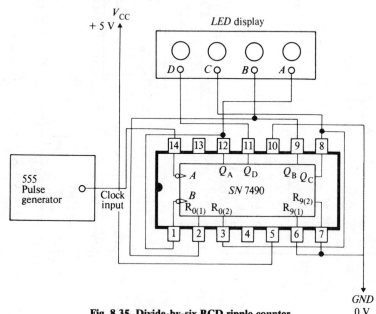

Fig. 8.35. Divide-by-six BCD ripple counter.

GND
0 V

124

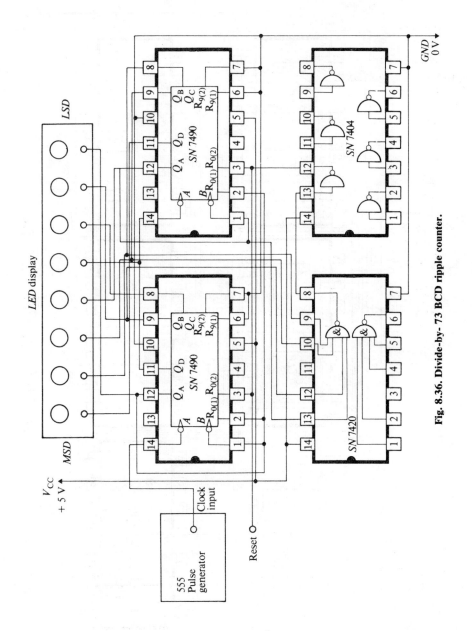

Fig. 8.36. Divide-by- 73 BCD ripple counter.

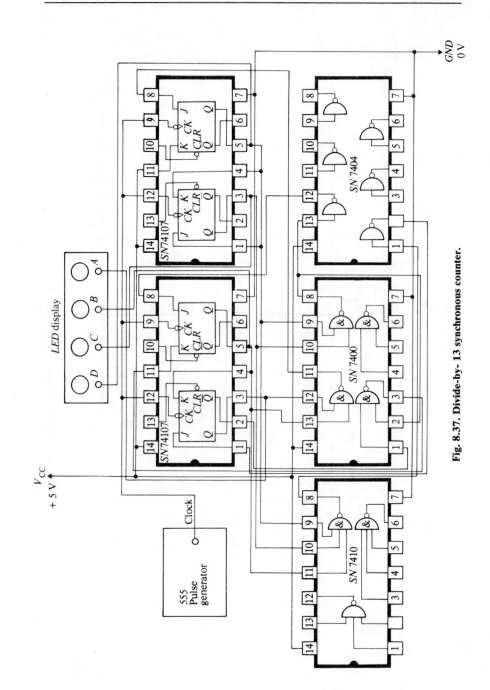

Fig. 8.37. Divide-by- 13 synchronous counter.

Apply a 4-bit byte to the input register, and the ADDRESS of the desired register to the SN 7442. Apply the ENABLE input and CLOCK pulses. Repeat this procedure by 'storing' other 4-bit bytes to different registers. *Read* the data stored in the registers on the LED's.

Practical exercise 8O

DIVIDE BY 6 RIPPLE COUNTER (BCD)

The SN 7490 Decade counter can be modified to change the count cycle by *decoding*. The decoding signal is fed back to the asynchronous CLEAR input to cause the counter to adopt a shortened cycle. The outputs which are in the logical 1 state at the end of the count are fed back to the RESET to 0 inputs ($R_{0(1)}$ and $R_{0(2)}$), as shown in Fig. 8.35, using $1 \times$ SN 7490 Decade counter.

Apply the CLOCK pulses, and observe the output on the LED's.

Practical exercise 8P

DIVIDE BY 73 RIPPLE COUNTER (BCD)

Larger division ratios may be achieved in BCD code by using two SN 7490's in cascade. BCD numbers with no more than two 1's in their sequence do not require additional external logic gates in the decoding network.

Connect up the arrangement of the counter shown in Fig. 8.36, using $2 \times$ SN 7490 Decade counter, $\frac{1}{2} \times$ SN 7420 Dual 4 i/p NAND gates, $\frac{1}{6} \times$ SN 7404 Hex inverter.

Apply the CLOCK pulses to the input and observe the output of the counter on the LED's.

Practical exercise 8Q

DIVIDE BY 13 SYNCHRONOUS COUNTER

Connect up the counter arrangement shown in Fig. 8.37, using $2 \times$ SN 74107 Dual J–K flip-flops, $\frac{1}{3} \times$ SN 7410 Triple 3 i/p NAND gates, $\frac{3}{4} \times$ SN 7400 Quad 2 i/p NAND gates, $\frac{1}{3} \times$ SN 7404 Hex inverter.

Apply a train of CLOCK pulses to the input, and observe the output on the LED's.

9 Industrial logic control systems

9.1 Introduction

Many industrial applications exist which depend upon a combination of input signals to initiate operation, such as the simple machine control example considered in chapter 4. However, a wide range of applications require a logic system which initiates operation after a particular sequence of events has occurred. These systems may use memory elements, counters and timers, as well as basic logic gates.

A simple example of these techniques is illustrated in the logic diagram of a combination lock shown in Fig. 9.1, in which the lock will be released when the push buttons are operated in the sequence C, A, B. Any other sequence will set off an alarm (audible or visible). The alarm may be cancelled, and the door may be relocked by simultaneously operating the push buttons A, B and C.

This chapter considers several applications using combinational and sequential logic circuits in which we shall describe the parameters of the system, and then examine the digital circuitry to achieve these objectives. The *practical* implementation of this digital circuitry to the particular industrial process would then be a further stage in the design procedure.

Practical exercise 9A

SIMPLE SECURITY SYSTEMS

Security systems are required in a wide range of applications from banks to the control of explosives used in site clearance. Assume that a storeroom (or shed) is used on a large site, and that it is required to control the access to (and movement of) explosives in storage. The shed is also used as an equipment store, so that a separate inner room is provided for the explosives. Three people have authority to use this storage, and each of them has a key. Now, any TWO keys will gain access to the equipment store, but ALL THREE keys are necessary to gain the further access to the explosive store.

These requirements may be satisfied with the logic network shown in Fig. 9.2.

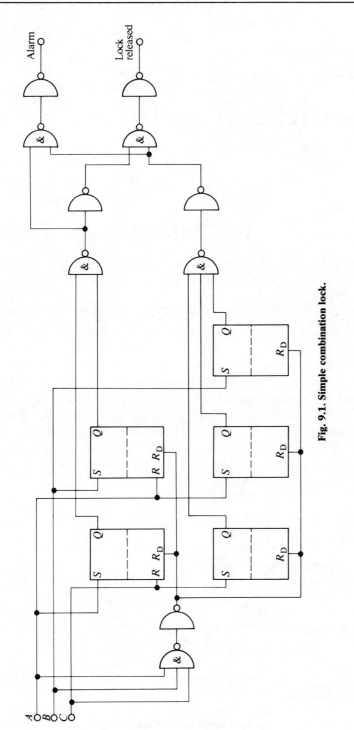

Fig. 9.1. Simple combination lock.

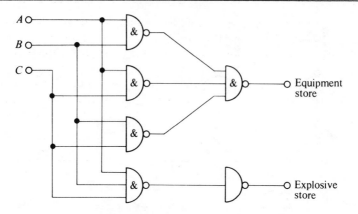

Fig. 9.2. Simple security system.

Connect up the simple arrangement shown in Fig. 9.2, using $\frac{3}{4} \times$ SN 7400 Quad 2 i/p NAND gates, $\frac{2}{3} \times$ SN 7410 Triple 3 i/p NAND gates, $\frac{1}{6} \times$ SN 7404 Hex inverter.

Apply all combinations of input signal—to simulate key operations—and monitor the output states on the LED's to verify the system operation.

Practical exercise 9B

PROCESS ALARM

Consider an industrial process in which it is desired to monitor the temperature in order to indicate the state of the process. When the temperature is normal, it is required to display a *GREEN* light. If a fault occurs such that the temperature goes high, it is required that the green light goes out and a *RED* warning light comes on. At the same time it is required to operate an audible alarm.

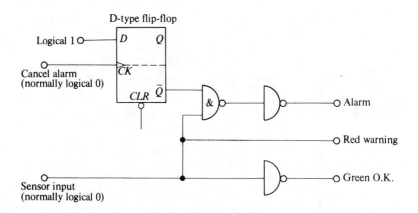

Fig. 9.3. Logic diagram of process alarm.

130

Provision must also be made for the operator to press a button to stop the alarm. Under these conditions the *RED* warning light must remain on until the fault has been cleared. The logic diagram for this system is shown in Fig. 9.3.

Connect up the network shown in Fig. 9.4, using $\frac{1}{2} \times$ SN 7474 dual D-type flip-flop, $\frac{1}{4} \times$ SN 7400 quad 2 i/p NAND gate, $\frac{1}{3} \times$ SN 7404 Hex inverter.

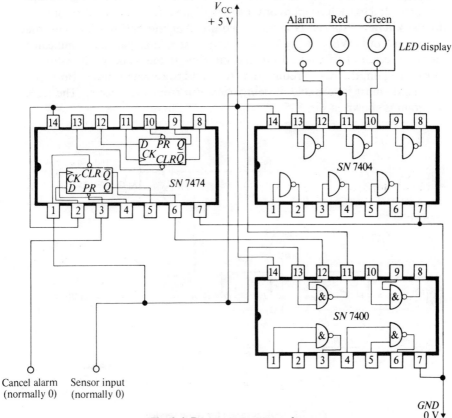

Fig. 9.4. Process temperature alarm.

Apply a logical 0 signal to the *CANCEL ALARM* push button and to the *SENSOR* input. Note that under these conditions the *GREEN* light is on.

Apply a logical 1 signal to the *SENSOR* input, and note that simultaneously the *GREEN* light goes out, the *RED* warning light comes on, and the *ALARM* is sounded.

Apply a logical 1 signal to the *CANCEL ALARM* push button and note that the audible *ALARM* is cancelled but the *RED* warning remains on.

131

Set the *SENSOR* input to logical 0 (fault cleared) and observe that the *GREEN* light comes on and the *RED* warning light simultaneously goes out.

Practical exercise 9C

HOTEL ROOM SERVICE

Consider a small private hotel or a large residence in which push buttons are provided in each room to operate a bell and a numbered indicator light in the kitchen. Normal conditions require that the bell is silent and the lamps are out. When a push button is pressed, that particular indicator lamp is required to come on in the kitchen. If the same push button is pressed again, the bell is required to ring, and to remain ringing. Both bell and lights must be capable of being cancelled from the kitchen. The logic diagram is shown in Fig. 9.5.

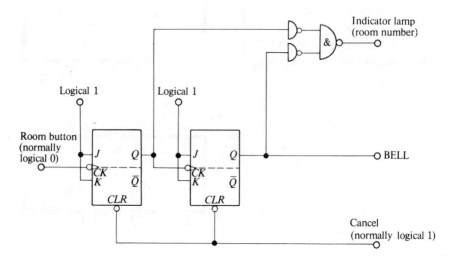

Fig. 9.5. Logic diagram for room service.

Connect up the network shown in Fig. 9.6, using $1 \times$ SN 74107 Dual J–K flip-flop, $\frac{1}{4} \times$ SN 7400 Quad 2 i/p NAND gates, $\frac{1}{3} \times$ SN 7404 Hex inverter.

Set the *ROOM BUTTON* to logical 1 and then to logical 0 and observe that the indicator lamp comes on.

Set the *ROOM BUTTON* to logical 1 and then to logical 0 a second time, and note that the *BELL* now rings and the light remains on. Set the *CANCEL* input to logical 1 and note that both the lamp and the bell cease to operate.

132

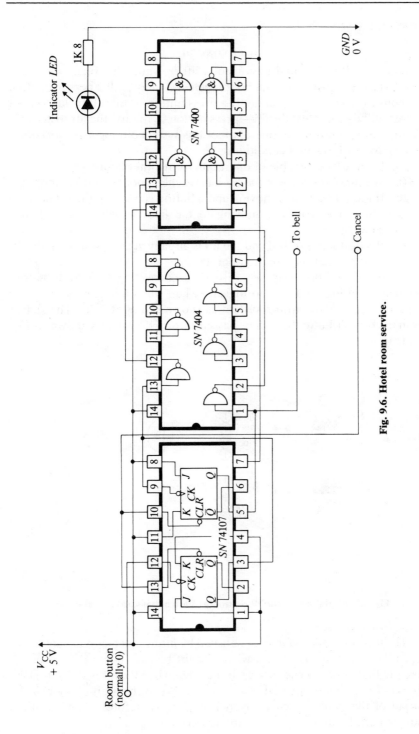

Fig. 9.6. Hotel room service.

Practical exercise 9D

LIFT CONTROL

Assume that it is required to control a lift operating between two floors. There are *two* buttons on each floor labelled ① and ②. If a person on floor ① requires to go to floor ②: firstly—press button ① to call the lift to floor ①; secondly—enter lift; thirdly—press button ② to raise the lift to floor ②.

In order to control the operation of this system our logic network must be able to satisfy several conditions:

(a) Lift must be capable of operation from either floor.
(b) We must know when the lift has reached floor ① and/or floor ②.
(c) If the lift is already moving and a button is pressed, the lift must be allowed to continue moving to the floor to which it is already moving.
(d) If a button is pressed and the lift is already at that floor, then the lift motor is not required to operate.

The first condition may be satisfied by connecting *both* push buttons labelled ① to logical 1 and to logical 0, i.e., when a button is pressed a logical 1 signal is fed to the system as shown in Fig. 9.7 (*a*). The signals from each push button are then passed to an OR gate, as shown in Fig. 9.7 (*b*).

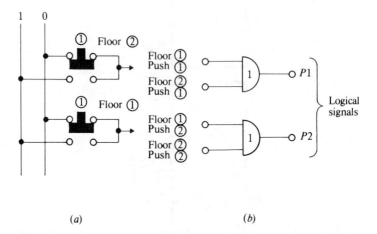

(a)   *(b)*

Fig. 9.7. Logical signals for lift control (*a*) Push buttons, (*b*) logical signals.

The second condition may be satisfied by fitting a detector at each floor, e.g., a limit switch to indicate when the lift reaches floor ① or floor ②. Assume that limit switches are used such that when the lift is at floor ① the output of that limit switch L1 is at logical 1. When the lift is at floor ② the output of that limit switch L2 is at logical 1. Therefore, when the lift is moving both L1 and L2 outputs are at logical 0.

Now, when the push button is operated, the lift is required to operate until it reaches the selected floor—when the limit switch is operated. This can therefore be controlled with a S–R flip-flop as shown in Fig. 9.8 (a).

However, the simple S–R flip-flop will not cater for the case when the lift is already at the selected floor, i.e., when $S = R = $ logical 1. This condition can be overcome by using the logic arrangement shown in Fig. 9.8 (b), which also prevents push button P1 (on either floor) from activating the LIFT DOWN flip-flop when the lift is moving towards floor ②.

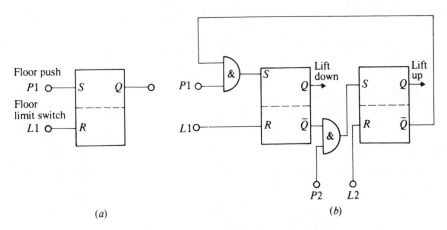

Fig. 9.8. Logical signals for lift motor drive control.

The complete system may be set up using J–K flip-flops as shown in the logic diagram in Fig. 9.9.

Connect up the arrangement shown in Fig. 9.10, using $1 \times$ SN 74107 Dual J–K flip-flops, $1 \times$ SN 7400 Quad 2 i/p NAND gates, $1 \times$ SN 7404 Hex inverter.

Set the CLOCK input signal from the 555 pulse generator to a fast speed. Apply input signals to simulate PUSH BUTTON operations and use LED's to indicate the logical state of the lift operation. Apply signals to simulate operation of the LIMIT SWITCHES—after the lift has been moving for a few seconds.

Practical exercise 9E

AUTOMATIC UNMANNED LEVEL CROSSING

In this case the requirement is to produce a warning signal in the form of flashing lights together with a bell to be initiated as a train *approaches* the crossing, and to continue until the train has safely passed the crossing.

The operating signals may be obtained by attaching a series of pressure switches at *three* positions along the railway lines. These pressure switches are used to detect the position of the train: the first being placed a safe

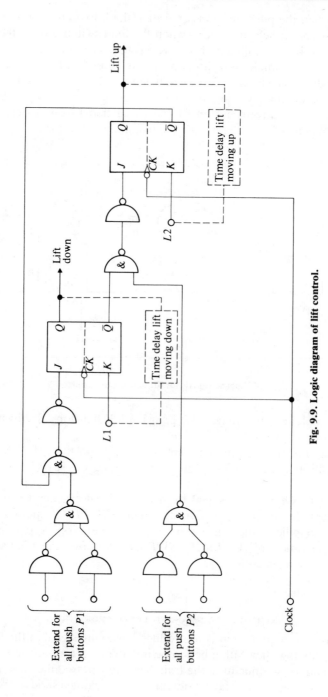

Fig. 9.9. Logic diagram of lift control.

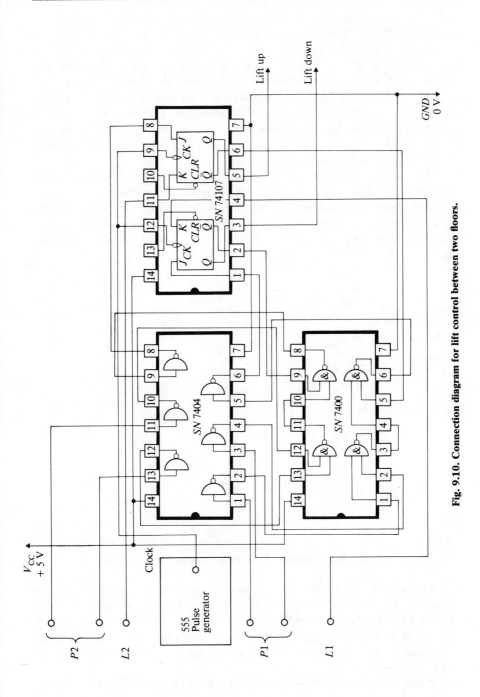

Fig. 9.10. Connection diagram for lift control between two floors.

distance *UP-line* (far enough from the crossing to allow adequate warning of approaching train); the second being placed *at* the crossing, and the third being placed a safe distance *DOWN-line* (far enough from the crossing to allow adequate time for the train to completely clear the crossing). The detector system needs to be duplicated for the UP-line.

Assume that the detectors are arranged so that as soon as the train moves over them a logical 1 signal is produced. It does not matter that whilst the train is moving over a particular switch, the output from that sensor will be (more or less continuously) logical 1—it is the *initial* signal in which we are interested for our system.

The outputs from the *three sensors* are applied to an OR gate, the output of which is applied to the CLOCK inputs of a *modulo 3 counter*, i.e., a simple binary counter which counts in the sequence 0, 1, 2, 0, 1, 2, etc., which is made up using two J–K flip-flops. The counter controls a S–R flip-flop to operate the audible/visible alarm, as shown in the logic diagram in Fig. 9.11.

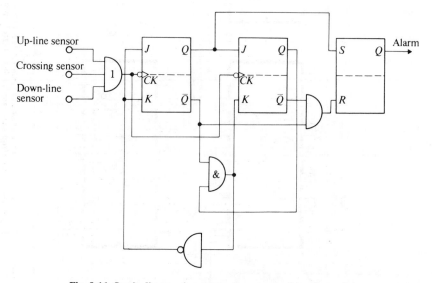

Fig. 9.11. Logic diagram for automatic unmanned level crossing.

Connect up the arrangement shown in Fig. 9.12, using $1 \times SN$ 74107 Dual J–K flip-flop, $\frac{1}{6} \times SN$ 74118 Hex S–R latch, $1 \times SN$ 7404 Hex inverter, $1 \times SN$ 7410 Triple 3 i/p NAND gate.

Apply a logical 1 signal to the *UP-LINE SENSOR* input, and then set this input to logical 0. Observe the state of the *ALARM* output using an LED indicator.

Apply a logical 1 to the *CROSSING SENSOR* input, and then set this input to logical 0. Check that the *ALARM* remains on. Finally, apply a logical 1 to the *DOWN-LINE SENSOR* input, and then set this input to

138

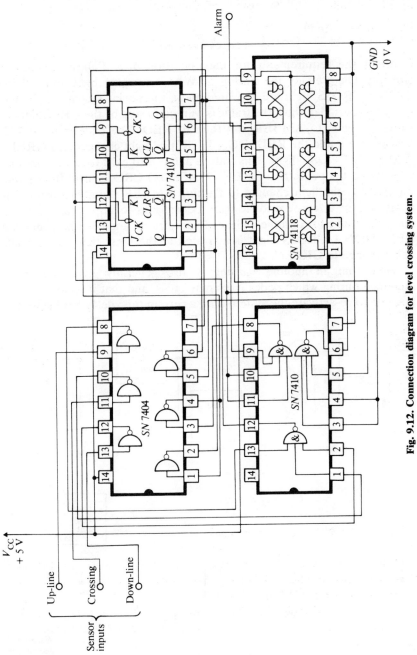

Fig. 9.12. Connection diagram for level crossing system.

logical 0. Check that the *ALARM* ceases to operate as soon as the *DOWN-LINE SENSOR* input is set to logical 1.

Practical exercise 9F

PELICAN CROSSING

Consider the requirements for the operation of a 'Pelican' pedestrian crossing. On pressing the button a 'WAIT' warning light and a RED 'DON'T CROSS' light are activated and a time delay system is initiated. After this period of time, the traffic is warned to stop by the GREEN traffic lights changing to AMBER and then to RED. The pedestrian 'WAIT' light goes out and the RED 'DON'T CROSS' light changes to a GREEN 'SAFE TO CROSS' light. This state is allowed to exist for a certain time, after which the pedestrian GREEN 'SAFE TO CROSS' light flashes (and the RED traffic light changes to flashing AMBER simultaneously). This continues for a few seconds to warn pedestrians to stop crossing, and to allow the traffic to continue if clear of pedestrians, and then the pedestrian RED 'DON'T CROSS' light is activated at the same time as the traffic lights change to GREEN.

Note that an audible warning is also generally given during the sequence during which pedestrians may cross, and that it is not normally possible to re-activate the system until a certain time delay has elapsed.

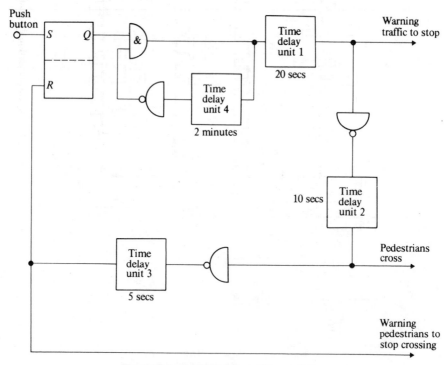

Fig. 9.13. Logic diagram for pelican crossing.

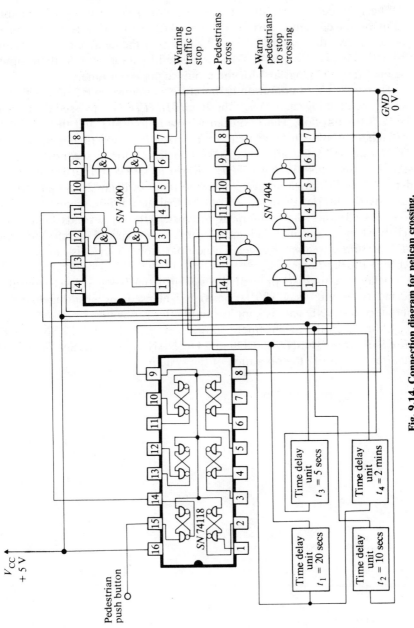

Fig. 9.14. Connection diagram for pelican crossing.

A simple logic network capable of dealing with the majority of these requirements is as shown in Fig. 9.13.

The time delay units shown could be produced by using monostable multivibrator elements, such as the SN 74121 with the external C and R chosen to give the required delay. Assume that the delay unit outputs are normally low, i.e., logical 0. When activated by a logical 1 at their input, the output goes to logical 1 for the requisite period of time.

The sequence of operation is therefore as follows: the S–R flip-flop is initially reset to logical 0. When the *PUSH BUTTON* is pressed the logical 1 SETS the flip-flop. The Q output is now at logical 1 and this is passed through an AND gate to activate time delay unit 1 and time delay unit 4. The output of delay unit 1 is thus at logical 1 for t_1 seconds (say, 20 seconds), and this *WARNS THE TRAFFIC TO STOP*. The output of delay 1 is fed through an inverter to the input of delay 2, so that after t_2 seconds the output of delay 2 goes to logical 1 for t_2 seconds (say 10 seconds) allowing the *pedestrians to cross*. After t_2 seconds delay 3 is activated and its output is at logical 1 for t_3 seconds (say 5 seconds) to *warn the pedestrians to stop crossing*. The output from delay 3 is also used to RESET the flip-flop.

Delay 4 is used to prevent the system from being re-cycled until a period t_4 seconds (say 120 seconds) has elapsed, by ensuring that the output of the AND gate is kept low until after t_4 seconds.

Connect up the arrangement shown in Fig. 9.14, using $\frac{1}{6} \times$ SN 74118, Hex S–R latch, $\frac{1}{4} \times$ SN 7400 Quad 2 i/p NAND gates, $\frac{2}{3} \times$ SN 7404 Hex inverter, and verify the operation of the system.

10 The digital computer

10.1 Introduction

Formerly, opinions of computers were divided into two extreme groups. Firstly, there were those who thought that the machine was almost God that knew and saw all things and could do everything short of making a good cup of tea. Secondly, there were those who were firmly convinced that the wretched thing would never work, and that (in one application for example) we should have to revert in the long run to that most flexible stock-recording and provisioning combination—the clerk, a pencil and a soft eraser! Those in the first group were the supreme optimists of this world, and for those in the second group we should not be too harsh, since there are still a few oddballs who are certain that the car will never replace the horse, and, let's face it, the Flat Earth Society is still with us—in spite of the many orbiting satellites and all the space flight activity during the last 20 years! Obviously, neither of these two groups is correct, since the computer is just a machine, made by man to extend his brain power in the same way that he uses other machines such as hydraulic lifts to extend his muscle power.

10.2 What is a computer?

The dictionary says that a digital computer is a calculator. 'So what!' you may say, 'so is an abacus, an adding machine, and you and I!' This is, of course, quite true, but there are THREE advantages that the computer has compared to ourselves:
   (a) very high speed,
   (b) very large capacity,
   (c) instant (or almost instant) recall from memory.
   In other words, the computer would not forget, and the same instructions would always produce the same results. Furthermore, the computer is not subject to sudden disappearances due to sickness in the middle of an important function, or in the habit of leaving cryptic notes such as 'back in 5 minutes' or 'gone for lunch'.

10.3 How does the computer work?

The computer really works quite simply—since the machine is an idiot! A very fast idiot maybe, but an idiot nevertheless! It is a machine designed to do *simple* arithmetic at high speed. The electronic innards may seem very complicated and costly—and they are! Also the telecommunications equipment associated with it may seem frightening, but the digital computer can basically do only ONE thing—and that is to ADD! When we tell it to add, multiply, subtract, divide, or make a simple decision—all that the thing can do is add! Furthermore, it is such an idiot that it can only add 1 and 1, and even then it does not make 2—but that is another story (binary arithmetic)! Hence the term 'digital'—it adds and extracts 'digits'. Oh yes, it is true that 'gimmicks' are built-in to enable it to take shortcuts, but to the layman it seems to go through a very laborious process in order to arrive at the answer. But, it IS FAST—and therein lies its power. Imagine trying to add a million a million times—the mind boggles! But, if you instruct a computer to multiply two numbers (by adding of course), then before you had even reached for your pencil, the computer would have given you the answer to that and numerous other problems—and the answers would be correct!

The machine can only do what it is told, and can only use the information that is fed to it. But, unlike the village idiot, it has a prodigious memory, is strictly logical and operates at very high speed. Thus, if we want to successfully operate the computer, there are certain basic priorities; firstly, any problems, work or information that we give the machine (input) must be put in extremely simple and precise terms. Secondly, any problem or work to be done must be based solely on the information we have already given it, or are about to give it, i.e., 'NEVER ASSUME', and, thirdly, the input must be made in a language that the machine can understand. Finally, we must tell it every step it has to take (program) in order to arrive at the required answer. We do the same thing with human beings by giving them courses (information), and job sheets (programs), but the results—to put the best possible light on it—are sometimes unfortunate.

10.4 Garbage in—garbage out (GIGO)

Now that we know, more or less how the computer works, you can appreciate that if we make a hash of the messages or information (input) we feed into it, we shall at best get the input rejected, and at worst end up knee deep in meaningless bumff in the flash of an eyelid! If we want to avoid such a disaster, it is essential to be accurate—hence the mnemonic GIGO!

For example, if the storeman on the shop floor *says* that he has issued ten items when in *fact* he has issued a hundred, as long as the computer record shows that there are ten or more items the machine would accept what he says. And why not? On the other hand, if he says that he has issued a hundred when the computer record shows that there are only ten, the machine would reject the input, since, stupid though it may be, even the computer knows that 100 from 10 is just not on! So again, be ACCURATE, since if your mistake is *logical* the machine will accept it, and the result could be highly amusing, embarrassing, or downright disastrous depending on your sense of humour. For example if you enter the wrong section code on a demand form, then provided the section and reference number demanded is in fact held in the computer's 'memory' you could end up with a 20 ton hydraulic press when all you wanted was an electric kettle for the section tea swindle!

Inaccuracies are also very costly. Suppose that a computer cost about a quarter of a million pounds to purchase and install, and a further thirty thousand a year to operate and maintain. Let us say that it takes 20 hours of every day to deal with the complete processing of all transactions and other requirements of a large organization. All well and good—FOUR hours a day spare! BUT, if it is assumed that 50% of the input must be rejected through finger trouble of one sort or another, it means that we would have to do another 10 hours of processing. Well it does not need me—or a computer—to tell you that science has not yet found a way of getting 30 hours out of one day! What is the solution? One answer would be to buy another computer at a cost of ————— and so on. We cannot afford this degree of carelessness.

10.5 Elements of the digital computer

A digital computer performs a sequence of arithmetic operations on data by means of a stored program of instructions, each of which define each step in the sequence.

A computer generally follows human behaviour when dealing with a problem. The original information, recorded in one way or another, is read into the machine—INPUT. The completed answer is read out of the machine—OUTPUT. In between these two is the 'work area' which we call the *CENTRAL PROCESSOR*, in which reference is made to the stored data and the stored program—*STORAGE*, and also in which the calculations are carried out—*ARITHMETIC*, and the whole process is supervised by the *CONTROL*.

The internal storage capacity (immediate access memory) of the central processor is not always sufficient to hold the program required and all of the data as well, so supplementary (back-up) external stores are used to which reference can be made in a short time.

A complete computer system therefore basically consists of a central processor unit (CPU) with a number of devices surrounding it called *peripherals*, as shown in Fig. 10.1.

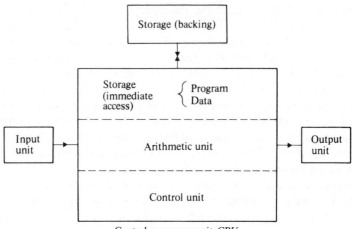

Fig. 10.1. Elements of the digital computer.

To operate a computer installation we must have a central processor and all the peripherals to support it—this is commonly referred to as the *HARDWARE* of the installation. In addition, we need the programs to direct the computer and the systems on which it will work—collectively known as the *SOFTWARE*.

10.6 The central processor

The central processor is the heart of the computer that deals with both *logic* functions and *arithmetic* functions together with the necessary control to carry out these tasks. Obviously, it will have input and output circuits and it will have a *memory* to store its instructions. It is usual for most function parts to have their own local temporary storage facilities which have several names but are generally referred to as *registers.*

The *memory* is required to store the instructions, and since these are needed over and over again, reading from this memory must not destroy the information. Using semiconductor memory this can take the form of *read only memory* (*ROM*). Temporary storage is used to store data when certain conditions occur, after which the data may no longer be required. In semiconductor terms this may take the form of *random access memory* (*RAM*).

10.7 The instruction

The organization of a computer depends to a large extent on the format of the program instruction. A single address format is shown in Fig. 10.2.

146

Operation code	Address

Fig. 10.2. Single adress format.

The *operation code* specifies the operation or sequence of operations to be carried out on data contained in a store location (address) specified by the number given in the address part of the instruction.

In a two address format, the operation code will involve two addresses, e.g., the instruction could be 'add the contents of one address to the contents of another'. A three address format would to the above, but may also specify a location for the result.

Space may also be provided in the format for any of the above cases for specifying a register, the contents of which could be used to modify the address part of the instruction. The purpose of this is to be able to use the same set of instructions to operate on data located at addresses other than those specified in the instructions by simply adding the register contents to the address part of each instruction as it is executed.

10.8 Storage

The purpose of the store may be outlined as follows:
- (a) to store the program,
- (b) to store the data,
- (c) to store any constants required for operations,
- (d) to store partial results of calculations,
- (e) to store final results prior to printing out.

The most important factors regarding a store are *speed* and *capacity*, and a compromise may be necessary for a given cost, e.g., the speed of calculations depends on how quickly numbers can be transferred between store and arithmetic unit. A fast access store is therefore desirable at some sacrifice of capacity—provided that data not immediately required may be held in slower but larger capacity storage. Therefore data required immediately for operations would be held in fast access storage such as magnetic core or semiconductor RAM's; the bulk of data would be held in slower bulk storage such as magnetic tape or disc. Data would be passed up in blocks to faster storage as the program dictates.

10.9 Arithmetic unit

This mainly consists of a number of registers connected by logic circuitry in such a manner to perform operations as dictated by the control unit. A register is effectively a store which contains a number or a group of numbers for as long as it is necessary to perform the specified operations.

The most important register is the *accumulator,* the purpose of which may best be illustrated by example, a *clear and add* instruction will set the contents of the accumulator to zero and insert a number from the address specified in the instruction. An *add* instruction will add a number to the contents of the accumulator so that it now contains the sum of the two numbers. This is the basis of most operations performed in the arithmetic unit; and other auxiliary registers may be incorporated as the particular operation demands.

10.10 Control unit

The essential parts of the control unit are:

(a) *The instruction register* which holds the instruction currently being executed.

(b) *The decoder* breaks down the operation code in the instruction into electrical signals which arrange the logic circuitry to perform that particular operation.

(c) *The address register* routes the address part of the instruction to the store where the specified location is interrogated for insertion or extraction of data.

(d) *The sequence control register* contains the address of the current instruction. When the current instruction has been obeyed, the contents of this register is raised by one. The contents now become the address of the next instruction. This is passed through the address register to obtain the next instruction from the store. This process assumes that each instruction is to be obeyed in sequence, but occasionally it is required to jump to another part of the program, either unconditionally or depending on a result obtained via the current instruction.

10.11 Input units

The *input unit* initially receives the program via some suitable medium, e.g., punched paper tape or cards, or teletype (TTY)—and then the data (via the same medium) can either be transferred to storage to wait to be operated on, or operated on directly as it is received in a real time system. There is no restriction on the type of input device, since the input unit acts as a buffer between the rate at which the device feeds in data, and the rate at which the computer is able to process it. Typical input devices include:

(a) *Paper tape reader*—paper tape, approximately 1 in. wide, with up to eight holes punched in a line across the paper, such a set of holes containing a code which is used to represent letters, numbers, and special characters. The code is read by photo-electric means—a typical reading speed is 1000 characters per second (with 10 characters per in. coded on the paper tape).

(b) *Card reader*—punched cards, approximately $7\frac{1}{2}$ in. $\times 3\frac{1}{4}$ in., each with 80 columns. Each column is used to represent one character by means of the holes punched. Reading is by photo-electric means at a typical speed of 800 cards per minute (i.e., about 1000 characters per second).

(c) *Teleprinter*–usually used as a remote terminal. The input may be either by pre-punched paper tape or by keying the required character. Typical speed 10 characters per second.

(d) *Console typewriter*—has limited application as a peripheral device, normally being used solely for the operator's control of the computer.

10.12 Output units

The *output unit* transmits processed data from the store to suitable output devices such as paper tape or card punches, line printers or transmission lines. This unit will buffer the rate at which data is transmitted via the computer, from the rate at which data is transmitted via the output device. Typical output devices include:

(a) *Paper tape punch* ⎫ both of these devices are normally restricted to pro-
(b) *Card punch* ⎬ viding output for use later as input.

(c) *Line printer*—capable of dealing with a large volume of output at high speed. The results are printed out a line at a time at a speed of up to 1000 lines per minute, each line can contain up to 160 character positions.

(d) *Teleprinter*—may be used as the major output if no line printer is available.

(e) *Console typewriter*—this has limited use as an output device, normally being used for messages to the operator from the program.

10.13 Computer applications

Computers are being applied to every branch of human activity. While some of these are not successful, there is no doubt that computers have found many far-reaching applications. Computer applications fall into three main areas:
  (a) scientific and technical calculations,
  (b) commercial projects,
  (c) military and space activity.
The branches of science where computers have been most successfully applied are those in which mathematics play an important part, e.g.,

astronomy, engineering, navigation, physics and chemistry. Scientific applications include: preparation of astronomical and navigational tables, calculation of artificial satellite orbits, design of nuclear power stations, processing of wind tunnel data for the design of aircraft, the analysis of stress recordings in aeroplane flight, X-ray photographic analysis and the design of new computers. M4C2N.

Commercial projects are applied in two clearly defined areas: office administration, and industrial processes. Computers are ideally suited for the calculation of wages, company accounting and the paperwork involved in stock control of large stores. Computers can be used to deal with a great deal of the work involved in production planning, material routing and progress state, e.g., preparation of management reports. Many specialized applications exist in financial organizations such as banking, insurance, the stock exchange, local and central government. Further special applications include transport, such as airlines and hotel reservations and the control of cargo movements. automation

In scientific and office work applications, the computer is used essentially as an arithmetical calculator and information store. In contrast, applications to industrial processes and to many military and space projects involve the use of the computer as a control system, i.e., the computer is directly connected to the process and initiates or modifies activity within it.

10.14   Computer developments

Digital computers have passed through a series of development stages, several of which are of particular interest to engineers and technicians. The introduction of the *MINICOMPUTER* a few years ago led to more economic design of computerised industrial control systems, resulting in a much wider application of computers for this purpose. More recently, the *MICROCOMPUTER* has hit the headlines. Generally, this takes the form of a *single chip IC microprocessor* (most, but not all, of them use MOS techniques) and the computer is completed by additional IC's for interfacing input, output and memory facilities. These peripheral IC's may be either MOS or TTL.

Many MSI and LSI functions are readily available using both techniques and microprocessors are generally made to be TTL compatible. Functions available in TTL include counters, registers, shift registers, storage registers, RAM's, ROM's, PROM's, multiplexers, encoders, decoders, etc.

11 Computer memories

11.1 Introduction

The *immediate access* memory in the computer CPU is made up of a matrix of high-speed random access storage cells in which any particular cell (i.e., the desired storage location) can be addressed by the co-ordinates on the X and Y axes of the matrix. In major computer installations, this requirement has been fulfilled by *ferrite core* magnetic storage elements. However, with the advent of minicomputer and microcomputer systems, together with the advances made in semiconductor technology, e.g., LSI techniques, many semiconductor memory elements are currently available and these are becoming much more widely used—particularly in the field of microcomputing. since the manufacturers of ferrite core memories adapt to the current market requirements, and their products are well-tried with a high reliability record, it is evident that not only will ferrite cores continue to be used in some large computer systems, but a trend is emerging where these manufacturers are producing memory systems for minicomputers and microcomputers as plug-in units.

The immediate access memory is connected to the CPU through a *buffer*, which is also used to rewrite data back into the memory if it is lost after a read operation.

Semiconductor memories include RAM's, ROM's, PROM's, EPROM's and EAROM's and may use any of the manufacturing techniques, e.g., bipolar, MOS, CMOS, NMOS, I²L, magnetic bubbles, charge coupled devices and more recently holographic memories.

Computer memory capability may be extended by using a slower bulk store such as magnetic tape, disc or drum. This is generally referred to as *backing storage* and the devices are usually considered as peripherals to the computer (CPU).

This chapter deals with the principles and essential techniques of many of the currently available computer memory systems.

11.2 Ferrite core store

The ferrite core store basically consists of matrices of ferrite cores threaded on wires. The cores are grouped to form *locations*, each location being a *computer word* (or *byte*) of a suitable size to store items of data or program instructions. The size of the word used depends upon the computer, but is generally in the range 8 to 48 bits (cores). The total storage capacity is generally quoted as a number of K words—where 1 K represents 1024 words (i.e., 2^{10}). Each core is approximately 0.4 mm diameter (0.015 in.).

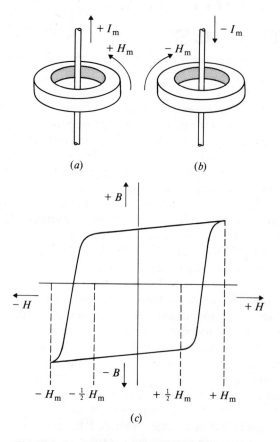

Fig. 11.1. Ferrite core magnetization. (*a*) Positive magnetization, (*b*) negative magnetization, (*c*) magnetic characteristic.

The ferrite core can be magnetized in either direction in terms of *two* remanent conditions of magnetization, by passing a steady current of I_m amperes through one wire and then removing it, as shown in Fig. 11.1 (i.e., either positively or negatively). Each core in the matrix has an X wire and a Y wire passing through it, so that it is possible to *select* any particular core by specifying the appropriate X and Y wires and passing *half* the current ($\frac{1}{2} I_m$) necessary to magnetise a core, as shown in Fig. 11.2.

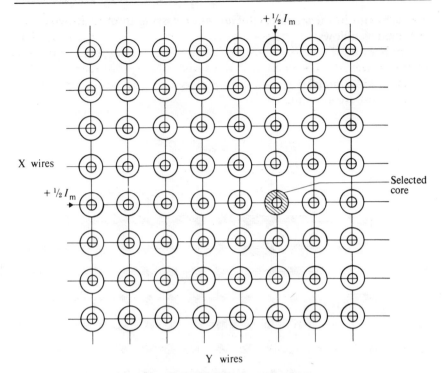

Fig. 11.2. Selection of a core in a matrix.

The matrices of cores are generally arranged in planes such that the corresponding core in each plane is used to represent a computer word. An additional plane is used for a *parity* check, as shown in Fig. 11.3, for an 8-bit word.

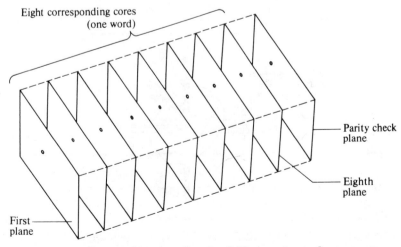

Fig. 11.3. Representation of an 8-bit computer word.

153

Each core in the matrix has *four* wires passing through it, which are identified as follows:

(a) X wire

(b) Y wire

(c) Z wire (inhibit)

(d) Sense (read) wire.

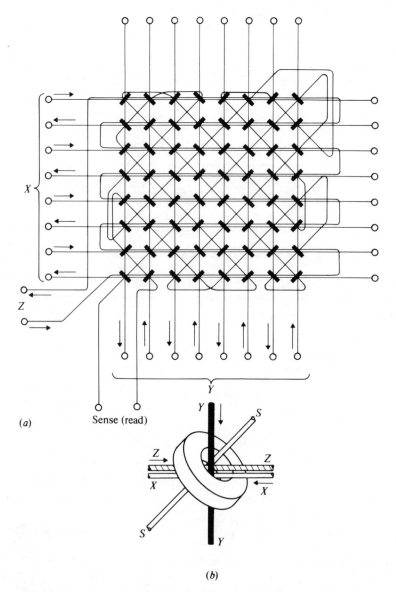

Fig. 11.4. Ferrite core matrix array with wires. (*a*) Matrix, (*b*) individual care.

In each matrix there is a *single* Z wire and a *single* read wire, both of which pass through every core in that matrix as shown in Fig. 11.4.

11.3 Reading and writing in store

When a current equal to half the value required to magnetize a core is applied simultaneously to the *selected* X and Y wires, the resulting effect is a current of sufficient value to magnetize the selected core at the inter-section of the two wires. The four operations which are performed in a core store are as follows:

(a) *Read cycle.* The state of a core is read by writing a logical 0 state to that core and detecting the change, if any, in its state. The read wire S, threaded through each core in the matrix is used to detect the change. If the state of the core changes, a current is induced in the S wire, i.e., if the original state was 0, there is no current, if the original state was 1, a current is induced.

(b) *Regenerative cycle.* After the 'read cycle' all the selected cores will be in the 0 state, so that if the information originally stored needs to be retained, it is necessary to regenerate that information in the cores.

  The selected X and Y wires have a reverse current passed through them of sufficient value to magnetize the appropriate cores to the 1 state. However, some of these cores may originally have been at logical 0. To prevent these cores being changed to 1 the Z (*inhibit*) wire is used, i.e., when a 0 is read a current is fed to the Z wire coincidentally with the half currents in the X and Y wires so that the resulting current is insufficient to change the core state from 0 to 1.

(c) *Write cycle.* This is a similar process to regeneration, with the Z wire current being derived from the information to be written. The write cycle is always preceded by the read cycle.

(d) *Parity check.* This is a method by which any transfer of data can be checked. *Odd-parity* is used to indicate that the total number of 1's in the word is always *odd* (including the parity bit). *Even-parity* is used to indicate that the total number of 1's in the word (including the parity bit) is always *even*.

During each cycle the parity of the word is checked to ensure that odd-parity (most common) is maintained—thus indicating a successful transfer.

  Core store is therefore a direct access system—each location has a numbered *address*. The time taken to access information (*access time*) is better than 2 μs and is the same for any location. This type of memory has a high reliability, low power consumption and occupies a relatively small space.

11.4 Thin film store

A thin film of nickel-iron alloy is deposited on glass in the presence of a strong magnetic field. This thin film then acts as a magnet which can be magnetized in either of two parallel directions to represent binary 0 and 1. The construction of the store is similar to that of the core store, except that conducting strips (on glass) are used instead of wires.

There is a one bit line for each row of thin film elements in each plane, and one *read* (sense) line for each plane. A *word* line is used to select the required element from each plane. The storage elements are sandwiched between the conducting strips during assembly.

Although using less power than core store and giving a faster access time (typically 0.2 μs), the output signal obtained requires a high degree of amplification.

11.5 Semiconductor store

A wide range of semiconductor memory elements are currently available, and although a standard design has not yet been established for a particular application, certain patterns in the form of packages and pin-outs are emerging as being compatible with other manufacturer's. The main requirements of semiconductor memories are that they should occupy a small area, have a fast access time and operate with low power dissipation. In addition, they should ideally be *non-volatile*, i.e., capable of retaining their information even when the power supply is removed. However, non-volatile stores really only apply to read only memories (ROM's).

Manufacturers need to decide two things: firstly, the process to be used, and secondly, the cell design. Bipolar techniques offer fast access times at low packing density, whereas MOS, and CMOS techniques give a slower access time at much higher packing densities. Memories differ from logic circuits in design since they consist of a regular pattern of basic cells—thus allowing a greater packing density than normal logic systems. The slower speed of unipolar based devices is mainly due to the parasitic capacitance associated the the bulk silicon substrate, and this is a problem which is gradually being overcome with advances in technology in the form of silicon on sapphire (SOS), NMOS, and more recently the new VMOS techniques (*AMI*).

All memory cells may be considered as either *static*, in which the stored information is maintained as long as the supply is on, or *dynamic*, in which the information is retained as a charge on a capacitor but must be periodically subjected to a *refresh cycle* to compensate for the leakage of charge from the capacitors.

11.6 Semiconductor static RAM's

The basic cell is a flip-flop, many of which are arranged in a matrix, in which any particular cell can be selected by X and Y select lines in a similar way to the ferrite core RAM. However, a *word* may be made up of, say, one row of cells in the array, which may be addressed simultaneously.

A simple bipolar static memory cell is shown in Fig. 11.5 (*a*). When the X and Y select lines are at a low voltage, current flows through transistor TR 1 or TR 2 (depending on the information stored) to the line which is at

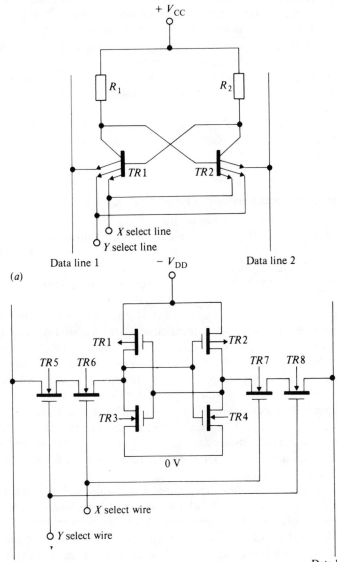

(*a*)

(*b*)

Fig. 11.5. Semiconductor static RAMs. (*a*) Bipolar, (*b*) unipolar (CMOS).

the low voltage. To select the cell, *both* X and Y select lines must be raised to a positive voltage, so that current is diverted from the select line to the data lines. The currents flowing in the two data lines will be unequal, giving an indication of cell state. To write in to a cell, the X and Y select lines must again be raised to a positive voltage, and the data lines held at a positive or negative potential to turn the desired transistor on.

A similar static cell using CMOS techniques is shown in Fig. 11.5 (*b*). Although this cell uses complementary MOS transistors, it is the n-channel devices which are used for addressing purposes—thus enabling much faster access times (typically 200 ns), and at much lower power consumption than bipolar techniques. The data lines are only connected to the memory cell when both the X and Y select lines are taken to a negative potential simultaneously.

Currently 4 K bit static RAM's are widely available in a single 22 pin or 16 pin DIL package. It is expected that single chip 16 K RAM will soon be readily available.

11.7 Semiconductor dynamic RAM's

Dynamic RAM's generally allow greater packing densities than static RAM's, and 16 K single chip dynamic RAM's are currently available with the promise of 32 K in the near future. These systems are faster, dissipate less power but suffer from the problem of leakage from the storage capacitor which means that it needs continuous refreshing (every few milliseconds).

Again, many possible circuits have been developed. One simple arrangement using three MOS transistors is shown in Fig. 11.6 (*a*), in which the charge is stored in capacitor C_1. If transistor TR 2 is turned on, this information may be refreshed (or altered). The cell may be *read* by connecting the read data line to a negative potential (logical 1), then TR 3 is turned on. If a logical 1 is stored in C_1, this turns TR 1 on which causes the data line to be discharged. If a logical 0 was stored in C_1, then the data line remains at logical 1. At the end of this read cycle, the data on the read line is the complement of the data stored in the cell. This is corrected by feeding the data back in through the write data line in order to refresh the cell after every read cycle.

Dynamic RAM's mainly use unipolar techniques, although bipolar systems have been used, and developments are still being undertaken with Schottky junctions and I^2L techniques. A simple bipolar dynamic RAM cell is shown in Fig. 11.6 (*b*), in which the data is stored in capacitors C_1 and C_2. Under normal conditions, the diodes are reverse biased which means that the data lines are isolated from the cell. The potential on one capacitor will be higher than the other depending on whether TR 1 or TR 2 is conducting. The cell is addressed by taking the word select line to a low potential. Information is now read by detecting the current flowing in

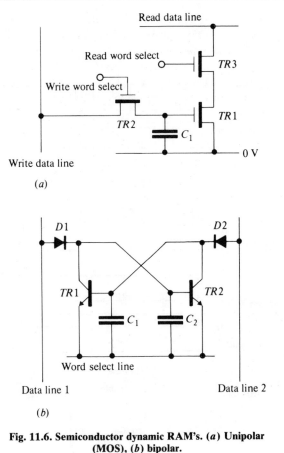

Fig. 11.6. Semiconductor dynamic RAM's. (a) Unipolar
(MOS), (b) bipolar.

the data lines, and the cell is refreshed (or written in) by forcing it into the desired state by applying a voltage to the data lines.

The memory cells are generally arranged in a matrix, and connected together with the X and Y lines to the X and Y Decoders as shown in Fig. 11.7 (a), in which each memory cell is assumed to be a three-transistor MOS arrangement as shown in Fig. 11.6 (a).

The amplifiers (shown in Fig. 11.7 (b)) are necessary to allow the cells to be refreshed every 2 ms or so. Transistors TR 3 and TR 4 are arranged to behave as resistances. The Data lines are changed to logical 1 via the precharge line, and then a row of cells are addressed. The output of these cells on their read data line is the complement of the cell content. The refresh cycle is initiated by turning TR 1 on to connect the complemented information to the write data line ready to be rewritten into the cell. Therefore a row of cells can be refreshed simultaneously although only one cell is being read. New information is written into a cell by keeping TR 1

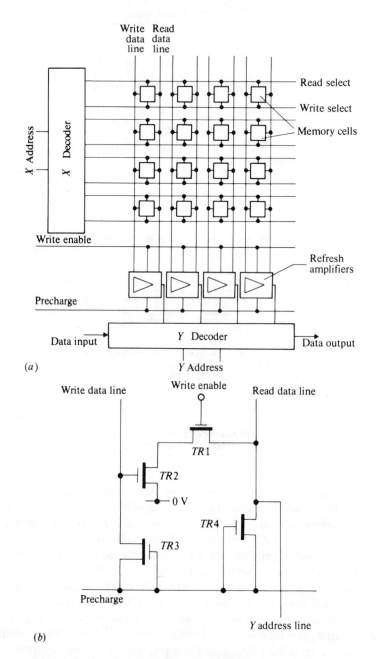

Fig. 11.7. Semiconductor dynamic RAM array. (*a*) **Array and inter-connection,** (*b*) **refresh amplifier.**

off and entering the information directly to the write data line. The total capacity of a semiconductor memory may easily be extended by using additional memory chips.

11.8   Semiconductor ROM's, PROM's, EPROM's and EAROM's

Although ROM's are usually made up in similar arrays to RAM's, so that any cell can be accessed as quickly as any other, ROM's are mainly required to store information which is not likely to be changed as often as it is in RAM's. Several different forms of ROM are currently available; the *factory programmed ROM*, which is *mask programmed* to a standard or customer specification, the *programmable ROM* (PROM) which is supplied in 'blank' form and the customer programs his own requirements—this form being ideal for development systems. There are two basic forms of PROM; the *fusible-link*, which, once programmed, is permanent, i.e., it becomes a ROM, and the *ultra-violet light erasable ROM* (EPROM) in which the program can be completely wiped clean by exposing to strong UV light, usually through a glass window in the package, then it is reprogrammed. Finally, the *electrically alterable ROM* (EAROM), sometimes referred to as the read mostly memory, in which data is erased by applying a high voltage pulse to the programming pins. An advantage of the EAROM over the EPROM is that a single word can be erased and rewritten without affecting the rest of the contents.

ROM's are essentially low cost, high speed, high density non-volatile storage matrices. Since they are generally required to store a predetermined pattern of logical signals, they can be basically made up of combinational logic networks instead of flip-flop elements. A simple diode matrix is shown in Fig. 11.8 (*a*), which is part of a denary to binary converter. The denary number is applied as a positive voltage to the appropriate address line and the output appears as a logical signal on the data lines. The diode pattern and interconnection is achieved by varying the masks used in the diffusion stages during manufacture. A transistor array, using MOS techniques is shown in Fig. 11.8 (*b*) which corresponds to the diode array. In this case, only *selected* transistor gates are connected to the address lines, so that a logical 1 signal input to an address line (i.e., a negative voltage) causes the transistors with their gates connected to be turned on, thus producing a logical 0 on its output data line. Mask-programmed ROM's are currently available in 16 K and 32 K bit packages.

The *fusible-link PROM* is available in two forms. In the first, a matrix of diodes (or transistors) is formed with a fusible link in series with each diode. Programming is achieved by fusing the links (with a controlled current pulse) in series with the diodes which are not required in the matrix, as shown in Fig. 11.9 (*a*). Once the required bit pattern is programmed (written into the memory) it cannot be changed. The second form is made up of a matrix as shown in Fig. 11.9 (*b*), in which each cell is

161

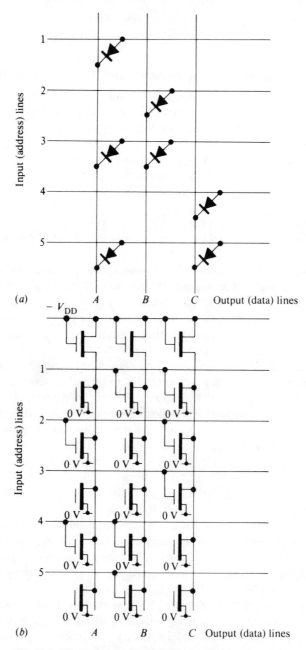

Fig. 11.8. Mask-programmed ROM's. (*a*) Output (data) lines,
(*b*) MOS transistor matrix.

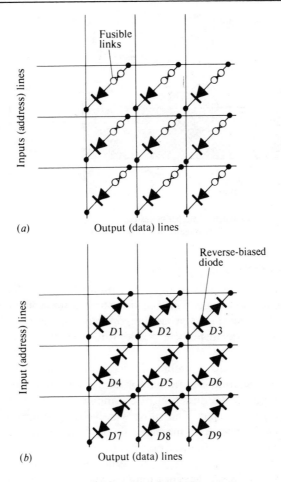

**Fig. 11.9. Fusible-link PROM's. (*a*) Fusible links,
(*b*) reverse-biased diodes.**

formed by two diodes connected back-to-back—thus making open-circuit cells initially. Programming is achieved in this case by applying a high enough voltage and passing a current pulse through the cells required, to break down the reverse biased diodes.

The *EPROM* can be completely erased by exposing it, through the window to strong UV light for about ten minutes. It is then reprogrammed in a similar way to that used for the fusible-link PROM. The EPROM operates by trapping a charge in a floating-avalanche-gate MOS transistor as shown in Fig. 11.10. EPROM's are currently available in 8 K bit chips. Standard EAROM's include packages of 32×16 bits, $2 \text{ K} \times 8$ bits and $1 \text{ K} \times 4$ bits.

163

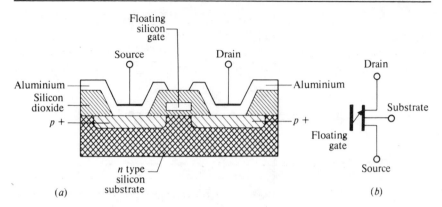

Fig. 11.10. Floating-avalanche-gate MOS transistor used in EPROM's. (a) constructional arrangement, (b) circuit symbol.

11.9 Backing storage

Since the immediate access storage is of finite size, it is often required to 'back-up' the CPU storage. This backing store caters for the high volume of data handled in most data processing operations, but has a slower access time. In general, the faster the access time the more expensive is the storage system.

Backing storage devices generally use a magnetic media on which to record data. Information is written to or read from a magnetic oxide film by means of changes in the direction of the magnetic field in the region of a magnetic read/write head. The most widely used system is *longitudinal recording*, in which the main magnetic field of surface magnetization is in the same plane as the oxide but parallel to it. A typical arrangement of a read/write head is shown in Fig. 11.11. When current is passed through the write coil the write head becomes magnetized and the magnetic flux links with the oxide film and causes it to become magnetized in the region of the

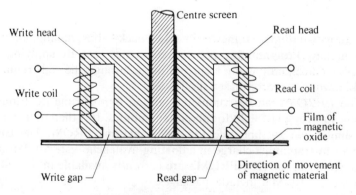

Fig. 11.11. Magnetic read/write head.

write gap. A current in the opposite direction would cause the induced magnetism to be of the opposite sense—these two states are used to represent binary 1 and 0. As the oxide film is moved past the read head, the magnetic flux in the region of the recorded areas cause an e.m.f. to be induced in the read coil.

The three most widely used backing stores are magnetic drum, magnetic tape, and magnetic disc (exchangeable disc storage).

11.10 Magnetic drum backing storage

The magnetic drum is a cylinder which is coated with a ferro-magnetic oxide, which is arranged to rotate at a constant speed about a vertical axis. The oxidized surface is divided into many horizontal tracks, i.e., the paths followed round the drum by each read/write head, and vertically into sectors, the boundaries of a sector being defined by the word size. Therefore, any word can be selected by the co-ordinates of the track and sector number. The arrangement of data on the drum surface is shown in Fig. 11.12 (a). If the data to be transferred is contained on more than one track, a switching delay from one track to the next may result, which could be as much as the time taken to complete one revolution of the drum. This delay may be minimized by staggering the start of the word in the next track. Three types of read/write assembly arrangements are possible:

(a) *Fixed head assembly*—one read/write head for each track. A fixed gap is finely set between the heads and the drum surface, and the

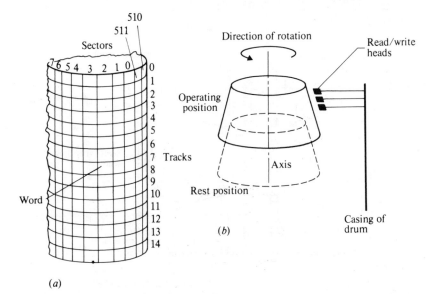

(a)

Fig. 11.12. Magnetic drum backing storage. (a) data layout on drum surface, (b) flying-head arrangement.

165

arrangement is housed in a constant temperature container to minimize the possibility of a head crash, i.e., a head coming into contact with the drum surface, due to expansion with heat.

(b) *Dancing head assembly*—one head for several tracks, positioned by an electromagnet. This system has a slower access time than the fixed head but is cheaper.

(c) *Flying head assembly*—one head for each track, aerodynamically shaped to 'fly' in the layer of moving air over the surface of the drum. The drum must be run up to the operating speed before the heads are in the correct operating position. A head-crash is minimized by using a slightly tapered drum which rises up its axis as it approaches its operating speed, as shown in Fig. 11.12 (*b*). This arrangement has a fast access time and allows a high packing density but is expensive.

11.11 Magnetic tape backing storage

Magnetic tape, similar to domestic recording tape but of much higher quality, and usually about $\frac{1}{2}$ in. wide, plastic coated on one side with a film of ferric oxide, is wound on reels holding 200, 600, 1200 or 2400 ft of tape. The tape is capable of storing up to 2000 characters per inch, and is assembled into a tape deck as shown in Fig. 11.13. The tape drive is effected by moving the pinch roller against the tape and onto the constantly revolving capstan. The tape is stopped by removing the pinch roller and applying brakes to the rape reel hubs. The tape drive must be capable of

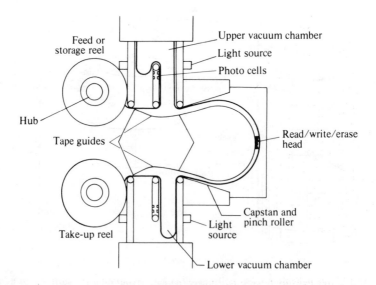

Fig. 11.13. Layout of magnetic tape deck.

responding to rapid start, stop, rewind and sudden runs without causing damage to the tape.

Information is stored on tape in *frames*, as shown in Fig. 11.14, in which a seven track tape is made up of six data bits and a parity bit. During reading the parity is checked and the parity bit is then not used again, and when data is written the parity bit is written and checked by a read-after-write operation. Data is transferred to and from the CPU in *blocks* which are separated along the tape by *inter-block gaps* which allows the tape to reach the required speed for read/write operations.

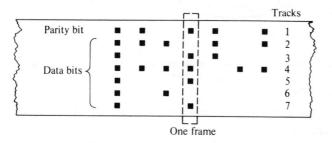

Fig. 11.14. Data recording on magnetic tape.

The head arrangement on the tape deck performs three operations, *erase, write* and *read.* During writing, the data previously recorded is erased, the new data is written serially in frames, a block at a time, and then checked by the read head. Access time is quite long (due to serial storage) and can be up to several minutes, but is relatively inexpensive.

11.12 Exchangeable disc storage (EDS)

The exchangeable disc store is made up of a cartridge of discs (typically six) mounted on a transport unit which includes the drive unit, read/write heads and the head assembly control circuitry. A cartridge of six discs, as shown in Fig. 11.15 (a), has ten recording surfaces on a ferro-magnetic oxide film—the outer two surfaces not being used. Each active surface is accessed by its own read/write head mounted on a metal arm. All the arms move together, with pairs extending between adjacent discs. When the cartridge has been run up to its operational speed the heads are moved to the centre of the cartridge, and then the individual heads are moved to their separate surfaces by rotation of the torsion arms which carry them.

Each active surface is made up of several hundred concentric *tracks*, and the recording area made up of the ten corresponding tracks (one from each disc surface) is referred to as a *cylinder*, as shown in Fig. 11.15 (b). Each track is divided into eight equal regions which are known as *blocks*, each of which has the same storage capacity. Data is written to and read

167

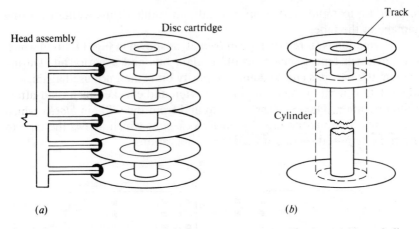

(a)   (b)

Fig. 11.15. Exchangeable disc storage arrangement. (a) Head assembly and disc cartridge, (b) formation of a cylinder.

serially within and between cylinders, and addressed in terms of the cylinder and head number.

EDS cartridges are available with up to 100 million characters per cartridge. Access time is fast at a maximum of 100 ms, but the cost is high. *Floppy-discs* have been developed to provide similar backing storage facilities for microcomputer systems. The principles are esssentially the same as those described for EDS but the discs are made of a more flexible material hence the name, and the discs and unit are much smaller than a mainframe peripheral.

12 Concepts of programming

12.1 Principles of flowcharting

Any problem which the computer is required to solve must be capable of being *written down as a solution* in a series of clearly defined steps—known as an *algorithm*. The algorithm for the solution of a problem by a machine is the specification of a finite number of instructions which, when executed by the machine, determines the actual solution (if, in fact, a solution is possible). This list of instructions may be represented by a diagram of interconnected symbols—known as a *flowchart*.

The *program flowchart* is a detailed description of the *program* to be used to solve a particular problem, and will invariably reflect the type of computer and the *language* to be used. There are many advantages in preparing a program in this way:

(a) It forces you to analyse the problem before you attempt to produce a solution.
(b) A clear description of how the problem is to be solved is presented.
(c) A record is provided which simplifies the task of finding errors in your solution.
(d) It can be used to describe to other people what has been done.

12.2 The simple flowchart

When a flowchart is being drawn, it is advisable to ask the following questions:

(a) What data is available? In what form is the data presented? In what units is the data measured? In what order is the data?
(b) What solutions are required? In what form are the solutions required? In what units are the solutions to be measured? In what order are the solutions required?
(c) What methods are available for the solution of this problem? Which of these methods is the most efficient?

The symbols most widely used in drawing flowcharts are shown in Fig. 12.1, together with an explanation of their meaning. We shall now consider

169

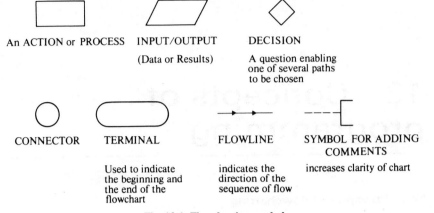

An ACTION or PROCESS   INPUT/OUTPUT   DECISION

(Data or Results)

A question enabling
one of several paths
to be chosen

CONNECTOR   TERMINAL   FLOWLINE   SYMBOL FOR ADDING
COMMENTS

Used to indicate   indicates the   increases clarity of chart
the beginning and   direction of the
the end of the   sequence of flow
flowchart

Fig. 12.1. Flowcharting symbols.

some simple examples to illustrate how these symbols are used in flowcharts.

Note: When *writing* flowchart and program data, ambiguity can be avoided by modifying some of the characters as shown below:

Numeral 'nought' is written as 0

Letter 'oh' is written as ϕ or θ

Numeral 'one' is written as 1

Letter 'eye' is written as I

Numeral 'two' is written as 2

Letter 'zed' is written as Ƶ

Numeral 'seven' is written as 7

Multiplication is written as *, and exponent as ** (this depends on the language).

Visual display units (VDU's) generally display 'nought' as ϕ, and 'oh' as 0.

Example 12.1

Draw a flowchart for the everyday example of 'telephoning a friend', assuming that the call is to be made from a telephone kiosk, and that only one operation can be performed at a time.

The solution is shown in Fig. 12.2.

Example 12.2

Draw a flowchart for 'getting up in the morning'.

The solution is shown in Fig. 12.3, from which it should be noted that there are many possible variations. The important points to appreciate are that *you must not take things for granted, i.e., NEVER ASSUME.*

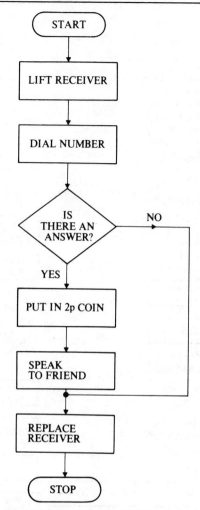

Fig. 12.2. Flowchart for telephoning a friend.

12.3 Arithmetic symbols

Although the two flowcharts considered above are for simple everyday tasks, many problems exist in which we are mainly concerned with arithmetic operations. The numbers specified in program flowcharts are generally referred to as *variables*—since their value can change. These variables— as in algebra—may be denoted by letters, or groups of letters, e.g., x, y, a, b, A, N, NUM, ANS, etc. Each variable occupies a storage location in the computer memory and is assigned values during the sequence of instructions specified by the flowchart. The *assignment* is usually denoted by one of the following:

$$= \qquad := \qquad \leftarrow$$

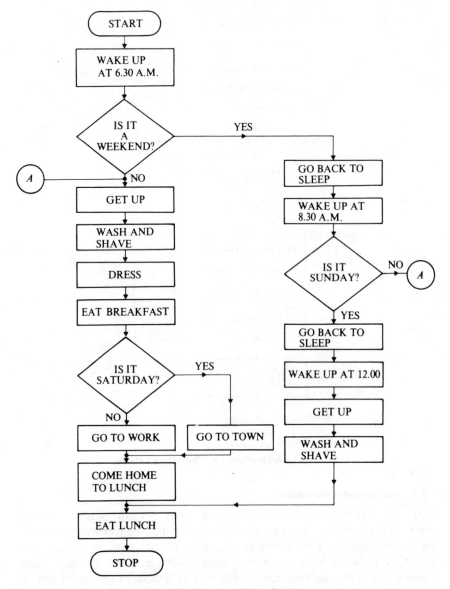

Fig. 12.3. Flowchart for getting-up.

For example, the algebraic statement $P = Q$ is used to denote the assignment of a value of the contents of store location Q to the store location P, *OR* location P takes the value of the contents of store location Q, *OR* $P \leftarrow Q$ is an alternative method of showing the same thing.

12.4 Arithmetic statements

When arithmetic operations involve the use of variables, we build up the arithmetic statement by writing:

$$\text{Variable} = \text{Arithmetic Expression}$$

where the arithmetic expression is made up of combinations of variables and standard arithmetic operations. For example $y = y + 1$, $x = y + z/a$, $x = (-b + \sqrt{b^2 - 4.a.c})/2.a$.

The complexity of the statement allowed is generally determined by the programming *language* which is to be used.

Decisions are all effected by comparisons, logical operations or arithmetic relationships. Some variations of the basic decision symbol are shown in Fig. 12.4.

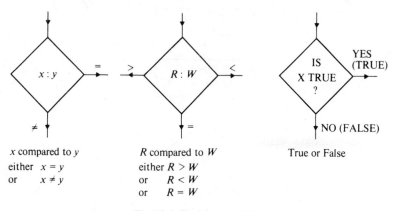

x compared to y
either x = y
or   x ≠ y

R compared to W
either R > W
or   R < W
or   R = W

True or False

Fig. 12.4. Decision symbols.

12.5 Looping

Looping is a process which enables the repeated use of a section of program. However, when preparing a program in which it is desirable to use looping techniques, it is essential to ensure that we can '*get out of the loop*'. This can be achieved in two ways:

(a) When we know how much data is being processed, i.e., when we know how many times we have to 'go round the loop', we can include a *counter* which is incremented by one each time we go round the loop. When the counter reaches the predetermined number, we exit from the loop to complete the remainder of the program.

(b) When we are processing an unknown quantity of data, we can add one data item (i.e., a data card) after all the program data. This card is coded with, say −1, or ****, and is called a *ROGUE VALUE*. The program looks for this rogue value every time it reads in data around the loop. Once the rogue value has been detected by the program we exit from the loop and continue with the remainder of the program.

Example 12.3

Draw a flowchart to raise Y to the power of N, where N is an integer.

The solution is shown in Fig. 12.5, in which a counter is used. When $M = N$, we exit from the loop.

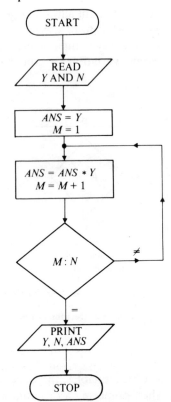

Fig. 12.5. Flowchart for Example 12.3.

Example 12.4

Draw a flowchart to solve the problem: given two *different* numbers A and B form a number C which is the sum of the largest squared plus the other.

174

Assume that all the numbers are positive, and that there is an unknown amount of data.

The solution is shown in Fig. 12.6, in which a rogue value of -1 is used.

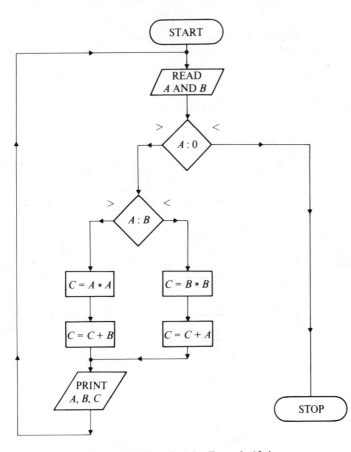

Figure 12.6. Flowchart for Example 12.4.

Example 12.5

Draw a flowchart to provide a solution for the following problem: given that P, Q, R, S and T are available as data input, it is required to compute A when the following conditions must be observed:

If $P = 2$, then $A = P^2 + Q + R - S$

If $P \neq 2$ and $Q = 3$, then $A = T^2 + Q + R - S$

If $P \neq 2$ and $Q \neq 3$ and $R = 4$, then $A = (P + T)^2 - Q - R + S$

Otherwise $A = +3$.

In this case, we will assume that there is *one* set of data only.

The solution is shown in Fig. 12.7, again assuming that only one operation may be executed at a time.

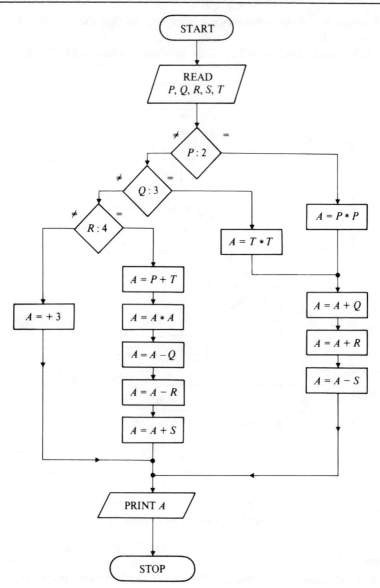

Fig. 12.7. Flowchart for Example 12.5.

Example 12.6

Draw up a flowchart suitable for the task of sorting English decimal coins into separate bags.

The solution is shown in Fig. 12.8.

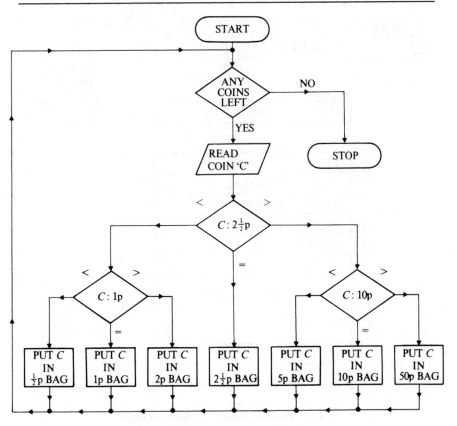

Fig. 12.8. Flowchart for Example 12.6.

12.6 A simple program

A *program* consists of a series of precise instructions to the machine. These instructions are loaded into consecutive 'pigeon holes' called *addresses* (or locations for words or bytes) in the computer store. The sequence control register scans these instructions in order and causes the computer to obey them. Arithmetical operations are carried out in the arithmetic unit, the results of operations appearing in a special register known as the *accumulator*. In general, movement of data in the computer, and to and from the computer, takes place through the accumulator.

A simple 'popular' example for the calculation of wages is shown in Fig. 12.9, in which it is assumed that each person's data is coded on to two cards, the first is the rate for the job and the second is the number of hours worked by that person. It should be noted that instructions would also generally be necessary that the data is being input via punched cards so a Card Reader is required, and a Line Printer is required to output the

177

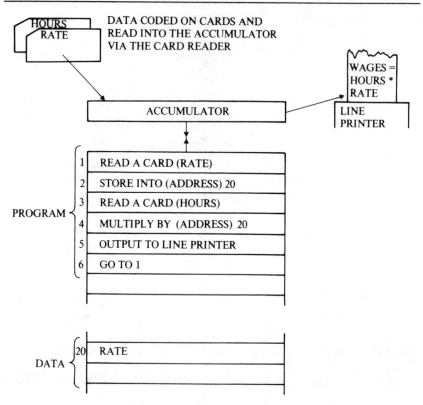

Fig. 12.9. A simple program 'calculation of wages'.

results. Furthermore, instructions must be included to stop the program, i.e., a counter may be included as described above for flowcharting.

12.7 Concepts of software

In any computer system, the software provides the interface between the human operator and the machine, and must ultimately result in a sequence of instructions being produced in a form which is acceptable to the machine. We have already experienced (chapter 2) some of the human difficulties of achieving this objective—even if the data is presented in binary, octal or hexadecimal, the human still has some difficulty in reading and understanding the coded patterns.

Various 'levels' of programming have now been established by one or more stages between the programmer and the machine acceptable code, these levels are: Machine Code, Assembly Language and High Level Language.

A program written in *machine code* consists of a list of instructions in binary form to be loaded into the computer memory for the computer to obey directly. It is therefore necessary to specify the number of the address of each word (byte) in memory whether it is instructions or data.

A typical instruction is 'add the contents of store location 50_{10} (binary 110010) to the contents of the accumulator, leaving the contents of store location 50_{10} unchanged!

In this case, the *operation* to be performed is ADD, and the *address* of the data to be operated on is 50_{10} or 110010_2. Assume that the code for ADD is 01, and assume that our word length is 8 bits, then this instruction will appear in store as:

$$\text{Operator} \quad \text{Address}$$

| 0 1 | 110010 |

Example 12.7

Write a program using the simple machine code listed below to solve the problem:

$$Q = P.U + \frac{Q.V}{R} - S.W$$

Machine code key:

Operation code N°	Mnemonic	Command	Meaning
01	CAD	Clear and add	Clear arithmetic section and add store location ——— to accumulator
02	ADD	Add	Add store location ——— to accumulator
03	STR	Store	Store accumulator in store location ———
04	SUB	Subtract	Subtract store location ——— from accumulator
05	MUL	Multiply	Multiply accumulator by store location ———
06	DIV	Divide	Divide accumulator by store location ———
07	PRT	Print	Print out accumulator
08	START	Start computer	Start computer, get the address of the first instruction word from the operation address
09	STØP	Stop computer	Stop the computer immediately

Assume that the values of the variables are in store locations as listed:

P is in store location 100
Q is in store location 101
R is in store location 102
S is in store location 103
U is in store location 200
V is in store location 201
W is in store location 202

Assuming that the instruction word length is 8 bits, then the program (with explanation) is as shown below:

Mnemonic	Instruction address	Instruction	Content of accumulator
START	000	08000001	0
CAD	001	01000200	U
MUL	002	05000100	P.U.
STR	003	03000300	P.U.
CAD	004	01000201	V
MUL	005	05000101	Q.V
DIV	006	06000102	Q.V/R
STR	007	03000301	Q.V/R
CAD	008	01000202	W
MUL	009	05000103	S.W
STR	010	03000302	S.W
CAD	011	01000300	P.U.
ADD	012	02000301	P.U+Q.V/R
SUB	013	04000302	P.U+Q.V/R−S.W
PRT	014	07000000	P.U+Q.V/R−S.W
STØP	015	09000000	

The process of writing machine code programs is obviously a laborious one, and is somewhat difficult. Furthermore, this type of program is time consuming and difficult to modify. However, machine code can be considered as being most appropriate to small dedicated systems, e.g., a microcomputer can be used to directly control an industrial process, in which the program instructions (software) can be stored in ROM, PROM, or EPROM and the data on which the computer acts will be measurement data made within the process and converted from analogue to digital form and fed directly into the microcomputer. This provides a relatively inexpensive system, since no additional peripherals are required for the computer.

An *assembler* is a special program which allows instructions to be written in the form ADD 50 or SUB TAX to be automatically translated into machine code, generally with one written mnemonic instruction corresponding to one machine instruction. Although programs may take a long time to write using these 'low level' languages, they usually result in very efficient programs in terms of store used and execution times.

A list of typical basic instructions for a mnemonic assembly language is given below, in which:

(A) means the contents of the accumulator

(n) means the contents of the store location n

(N) denotes an integer N (assumed to be positive and within the range 0 to 999).

180

Instruction	Operation	Comments
LDA n	$(n) \to A$	(n) unchanged
STA n	$(A) \to n$	(A) unchanged
ADD n	$(A) + (n) \to A$	(n) unchanged
SUB n	$(A) - (n) \to A$	(n) unchanged
MLT n	$(A) * (n) \to A$	(n) unchanged
DIV n	$(A)/(n) \to A$	(n) unchanged
LDAN	$N \to A$	This range of
ADDN	$(A) + (N) \to A$	instruction deals
SUBN	$(A) - (N) \to A$	directly with positive
MLTN	$(A) * (N) \to A$	integers and not
DIVN	$(A)/(N) \to A$	with store locations

Example 12.8

Using the above instructions, write the instructions necessary to perform the following:

Assuming x is stored in location 12, compute $(x + 3)$. 40 and store the result in location 12.

Solution:

> LDA 12
> ADDN 3
> MLTN 40
> STA 12

Example 12.9

Write the instructions to perform the following operations:

Add the two numbers stored in locations 25 and 26, store the result in location 25 and zeroize location 26.

Solution:

> LDA   25
> ADD   26
> STA   25
> LDAN 0
> STA   26

Most assembly languages have many more instructions which include versatile functions such as 'jump', 'modification' and 'function' instructions which allow more complex operations to be performed in response to relatively simple written (mnemonic) instructions. One commonly used assembly language is the *ASCII* code (American Standard Code for Information Interchange), which is commonly referred to as ASK–EE.

High level languages are completely independent of the machine, relatively easy to learn and allow the programmer to concentrate on the problem to be programmed. There are two types of high level language, the *interpretive* such as *BASIC* (Beginners All-purpose Symbolic Instruction Code), which is often referred to as a conversational language since the form of instructions and statements are more humanly biased. This

181

type of language is translated into machine code by means of an *interpreter*. The second type of high level language is converted into assembly language by means of a *compiler* before final translation of assembly into machine code. The compiler is a program containing a list of statements used in the problem-orientated language, and for each statement a list of machine instructions necessary to perform that statement. Then, by running the source (problem-orientated) program with the compiler program, an object (machine code) program is produced, which is then used to process the data. During compilation, as each source program instruction is read in, the compiler scans it for errors in the construction of the statements, and gives instructions for these errors to be printed out—the error messages being termed *diagnostics*, which is of great assistance in *debugging* (finding faults in the program and correcting them). Logical errors will not be revealed in this process, only the errors which the machine cannot recognise, i.e. the *form* of the instructions.

Many high-level languages have been developed, some of which are:

*FORTRAN—FOR*mula *TRAN*slator
*COBOL—CO*mmon *B*usiness *O*rientated *L*anguage
*ALGOL—ALGO*rithmic *L*anguage
*CORAL—C*omputer *O*n-line *R*eal-time *A*pplications *L*anguage. This is a
development of ALGOL. CORAL 66 was originally developed by the RRE, Malvern for military projects, but has since been adapted for such applications as Airline Reservations and the Post Office Viewdata system.

12.8 Preparing the program

We have examined briefly the principles of flowcharting, enabling a complete breakdown of the problem, and which can now be used in writing the program to be presented to the machine. When relatively simple tasks are being performed using machine code, the program can be 'written in', one instruction at a time, by setting switches on the front panel. Although this may be useful in a microcomputer prototype development system, it becomes laborious for anything but the simplest program and is therefore prohibitive.

Programs are therefore hand written onto program data sheets, and then punched on paper tape or cards for subsequent reading in by the appropriate peripheral, or the program can be 'written in' using a teletypewriter (TTY), a keyboard display unit (KDU), or a visual display unit (VDU).

The VDU provides very useful 'terminal' equipment, allowing the operator to 'write in' to the computer via a keyboard (similar to that of a typewriter) and giving a 'read out' on the screen (similar to that of a television), so that programs can be written and solved very quickly, the main disadvantage being that a hard copy of the program is not produced.

Appendices

A Logic symbols

Many different systems of symbolization have been used in equipment drawings and publications throughout the world. Early British Standard Specifications used circles for all the basic gates, with inhibition shown by a short perpendicular line drawn across the data flow line. The revision, in 1969, of BS 3939 Section 21—Logic Symbols, led to the D outline symbols being widely used in the United Kingdom and particularly in education. This Standard was revised in July 1977, and recommends the use of

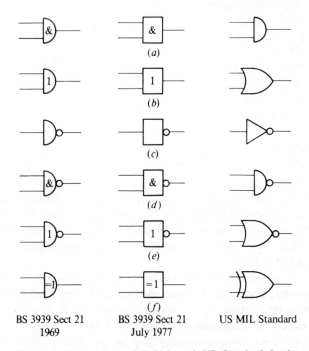

| BS 3939 Sect 21 | BS 3939 Sect 21 | US MIL Standard |
| 1969 | July 1977 | |

Fig. A.1. Comparison of British and US Standard Logic Symbols. (*a*) AND function, (*b*) OR function, (*c*) NOT function, (*d*) NAND function, (*e*) NOR function, (*f*) Exclusive—OR function.

rectangular outlines. However, due to the lead that the United States of America have established in the component and equipment manufacture and application, the US Military Standard has been the most widely used throughout the world. It is expected that this trend will continue until a European Standard is established and accepted, so that we shall encounter different 'standards' for some time to come. A comparison between some of the most common symbols is shown in Fig. A.1.

B Pin-out connections of popular range of 74 Series TTL IC's

The pin-out connections are shown in Fig. B1 for a range of 74 Series TTL IC's. The pin connections are always numbered *as viewed from the top* of the package. The list is by no means conclusive—most manufacturers and/or distributors issue complete lists of their own products. The IC numbers have now evolved as standard products throughout the industry, but it is always advisable to check with the particular manufacturer's data if in doubt.

Some of the commonly used abbreviations on the pin-outs are listed below, together with their meanings:

NC	Not connected
LT	Lamp test
CK	Clock
RBI	Ripple blank input
RBO	Ripple blank output
R_0	Reset to 0
R_9	Reset to 9
G	Enable
PR	Preset
CLR	Clear
Σ	Sum

When using logic gates, we must not have any 'floating' inputs. For NAND gates, all unused inputs must be connected to logic 1, and for NOR gates, all unused inputs must be connected to logic 0—or, connect the inputs together. We must not connect gate outputs together, but only *through* another gate.

Most TTL device inputs are 'enabled' by the application of a logic 1 signal. However, some inputs, such as PRESET and CLEAR inputs on flip-flops, may be enabled by the application of a logic 0. These are easily identified on pin-out diagrams as shown in Fig. B.2 (*b*).

Clock inputs are always identified by an arrowhead as shown in Fig. B.2 (*a*), where the device changes state on the leading edge of the clock pulse. Devices which are triggered on the trailing edge of the clock pulse, e.g., master-slave flip-flops, are identified as shown in Fig. B.2 (*c*).

184

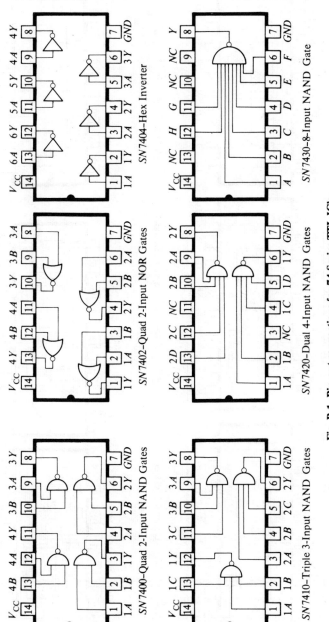

Fig. B.1. Pin-out connections for 74 Series TTL IC's.

SN7404–Hex Inverter

SN7402–Quad 2-Input NOR Gates

SN7400–Quad 2-Input NAND Gates

SN7430–8-Input NAND Gate

SN7420–Dual 4-Input NAND Gates

SN7410–Triple 3-Input NAND Gates

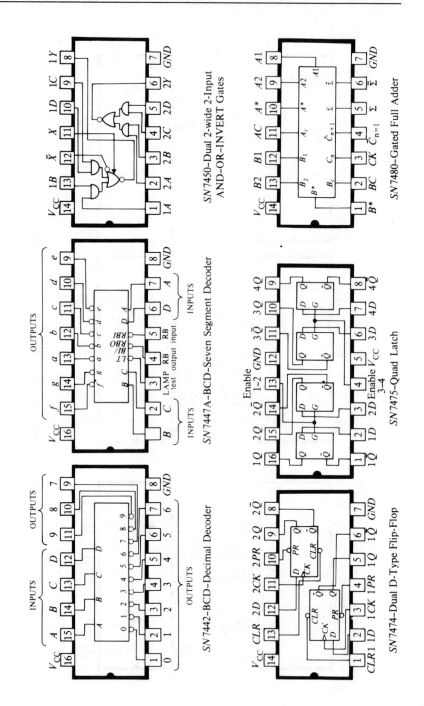

SN7450–Dual 2-wide 2-Input AND–OR–INVERT Gates

SN7480–Gated Full Adder

SN7447A–BCD–Seven Segment Decoder

SN7475–Quad Latch

SN7442–BCD–Decimal Decoder

SN7474–Dual D-Type Flip-Flop

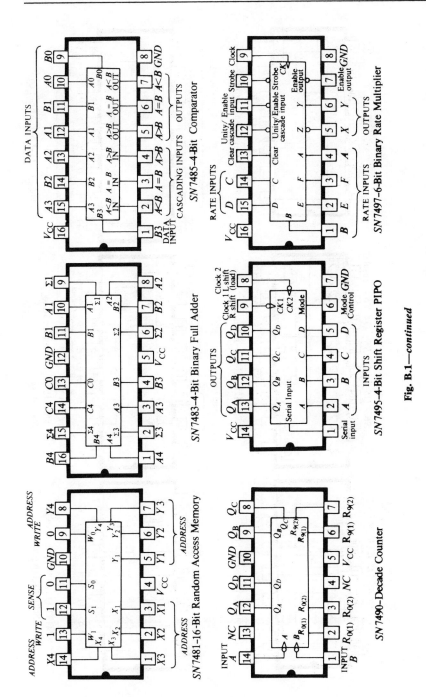

Fig. B.1—*continued*

187

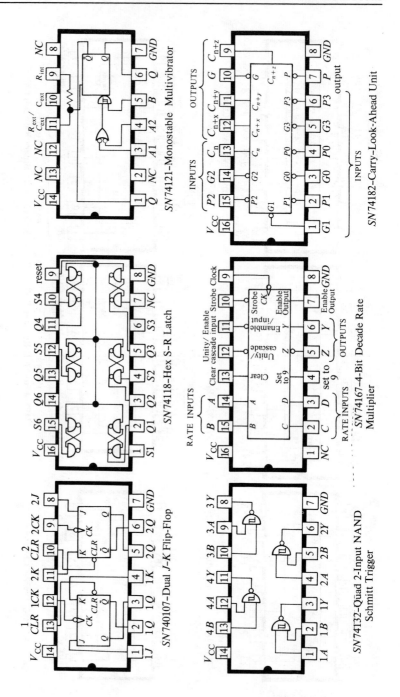

SN74121–Monostable Multivibrator

SN74182–Carry-Look-Ahead Unit

SN74118–Hex S–R Latch

SN74167–4-Bit Decade Rate Multiplier

SN740107–Dual *J-K* Flip-Flop

SN74132–Quad 2-Input NAND Schmitt Trigger

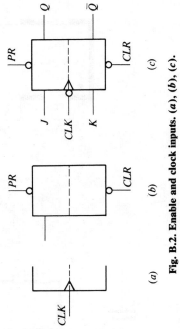

Fig. B.1—*continued*

Fig. B.2. Enable and clock inputs. (*a*), (*b*), (*c*).

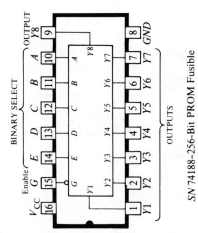

SN 74188–256-Bit PROM Fusible

C The 555 pulse generator and other useful arrangements

(a) *The 555 pulse generator*—The 555 timer is a highly stable controller capable of producing accurate time delays, or oscillations. The pin-out for the DIL version is shown in Fig. C.1 (*a*), and the circuit for a variable frequency (approximately 1 Hz to 10 Hz) Pulse Generator is shown in Fig. C.1 (*b*), in which the control is provided by the 1 M0 variable resistor. The frequency range could be changed by using a different value of capacitor.

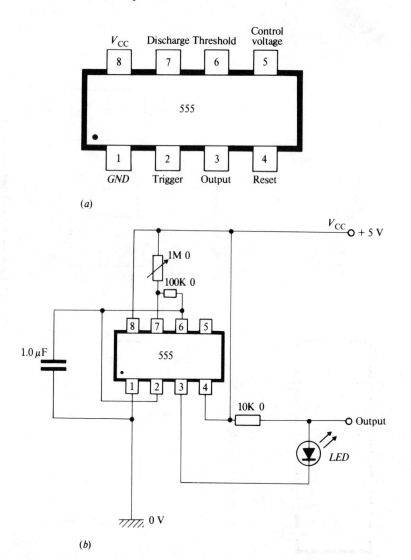

(*a*)

(*b*)

Fig. C. 1. The 555 pulse generator, (*a*) Pin-out for 555 timer, (*b*) circuit arrangement for pulse generator.

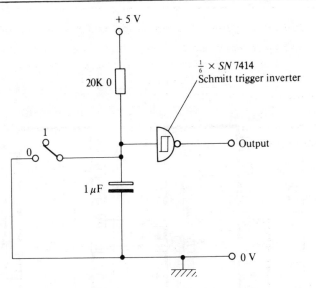

Fig. C.2. Switch-bounce eliminator.

(b) *Switch-bounce eliminator*—When mechanical switches are used to provide logical signal inputs, 'contact bounce' (as the contacts close) can cause problems, e.g., it can cause the input to become momentarily floating for the duration of the bounce. This effect can produce several output pulses from the switch where only one is required. This may appear to be quite satisfactory when setting d.c. levels—such as testing the logical operation of a logic gate, but causes errors if the switched pulses are being counted. Thus, counter inputs must be clean

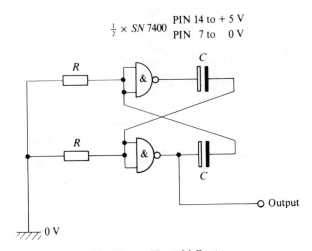

Fig. C.3. Astable multivibrator.

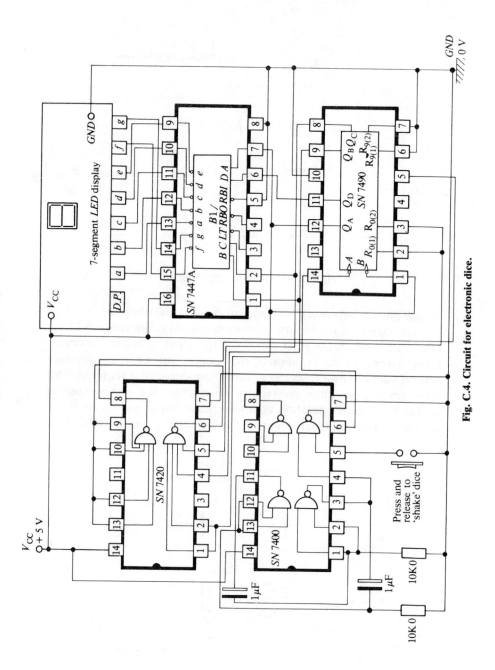

Fig. C.4. Circuit for electronic dice.

sharp-edged pulses, and in such cases a mechanical switch must be *de-bounced*. A switch-bounce eliminator arrangement is shown in Fig. C.2, using one Schmitt input inverter.

(c) *Astable multivibrator*—An inexpensive method for generating pulses to drive 'clock' inputs in digital IC's is shown in Fig. C.3, using two NAND gates, two capacitors and two resistors. Typical values for low frequency pulses are 470–1000 μF and 470R0–2K2, and for high frequency pulses 220 nF–1 μF and 470R0–10K0.

(d) *Electronic dice*—Many arrangements are possible for an electronic dice utilizing the astable multivibrator above. One simple method is shown in Fig. C.4, in which the astable multivibrator generates squares waves at about 1 kHz. This 'clock' signal is fed into the SN 7490 decade counter when the 'shake dice' switch is depressed—and disconnected when the switch is released, thus feeding in an unknown number of pulses. The counter is caused to cycle through a count of 1 to 6 inclusive by using a decoding network to reset to 0 after a count of 6 and connecting the ripple blank input (pin 5 on the SN 7447A) to 0 V, to eliminate the 0 from the display.

Index